Functionality and Protein Structure

Functionality and Protein Structure

((

Akiva Pour-El, EDITOR
PEACO

Based on a symposium
sponsored by the ACS
Division of Agriculture
and Food Chemistry at
the 175th Meeting of the
American Chemical Society,
Anaheim, California,
March 13, 1978.

ACS SYMPOSIUM SERIES **92**

AMERICAN CHEMICAL SOCIETY

WASHINGTON, D. C.   1979

Library of Congress CIP Data

Functionality and protein structure.

(ACS symposium series; 92 ISSN 0097-6156)

Includes bibliographies and index.
1. Proteins in human nutrition—Congresses. 2. Food
—Sensory evaluation—Congresses. 3. Food—Analysis
—Congresses.
I. Pour-El, Akiva, 1925-   . II. American Chemical
Society. Division of Agriculture and Food Chemistry.
III. Series: American Chemical Society. ACS sympo-
sium series; 92.

TX553.P7F86   612'.398   78-31964
ISBN 0-8412-0478-0   ASCMC 8   92 1–243 1979

S.D. 4/27/79
PSA

ACS Symposium Series

Robert F. Gould, *Editor*

FOREWORD

The ACS SYMPOSIUM SERIES was founded in 1974 to provide
a medium for publishing symposia quickly in book form. The
format of the SERIES parallels that of the continuing ADVANCES
IN CHEMISTRY SERIES except that in order to save time the
papers are not typeset but are reproduced as they are submitted by the authors in camera-ready form. As a further
means of saving time, the papers are not edited or reviewed
except by the symposium chairman, who becomes editor of
the book. Papers published in the ACS SYMPOSIUM SERIES
are original contributions not published elsewhere in whole or
major part and include reports of research as well as reviews
since symposia may embrace both types of presentation.

CONTENTS

PREFACE

Terminology and Classification

The term functionality is of recent origin. It was coined about 15 years ago by scientists dealing in plant protein utilizations. Its application since then has been widened, not always with adequate precision and consistency. Recently (1), it has been redefined as follows: functionality is any property of a substance, besides its nutritional ones, that affects its utilization. Although at first glance this definition is too broad, after careful examination it is the only one that encompasses the various areas of functional investigations now in progress.

The terms functionality or functional property will be used synonymously. Functional methods or functional evaluations refer to the procedures involved in testing these properties. Functionality investigations designate deeper evaluations of the properties, mostly connected with their relation to other physicochemical characteristics of the substances studied. These terms, although not used with rigor up to now, are put forth in this manner for possible standardization of reports in this area in the future.

In proteins, functionality investigations can be roughly divided into two main areas, depending on utility: (1) Investigations of properties affecting the utilization as a food or food additive; and (2) investigations of properties affecting the enzymatic activity of the product. The first class sometimes is also subdivided between the sensory evaluations and those using other physicochemical measurements.

Significance of Protein Functionality

The need for supplying an expanding world population with adequate protein foods will not be discussed here. It has become a common enough belief, regularly reiterated, and widely documented in the literature of protein investigations. It is, however, necessary to dwell briefly on the status of protein functionality investigations in the total area of protein studies.

The enzymatic branch of protein functional investigations is significant, not only for the basic scientific knowledge it imparts which helps clarify biological processes, but also because, by elucidating how catalytic functions depend on specific protein structures, more useful and more powerful enzymatic processes might be developed for food production and other industrial needs. Highly specific enzyme activity is always in demand for elimination of side reactions. This could be the outcome of a more complete understanding of the relation between enzymatic activity and protein structure.

In the food area, protein supplies are emphasized more frequently and are mostly studied for the nutritional properties. It should be pointed out, in the strongest terms, that protein foods are rarely used as crude powders or in their native forms. They are ingested most frequently as part of a complex food system where their functionality, rather than their nutrition, is the property most obvious to the consumer. In fact, many projects to alleviate protein malnutrition in less-developed countries have floundered because the introduced food forms did not fit the accepted pattern (i.e., the functionality) of the foods normally used in the region. Therefore, it is now commonly recognized, for improved nutritional standards, that any new food introduced into a population must, of necessity, be considered for its functional properties. Improving these properties will be a major factor in the successful adoption of the "new" food by the people in the area. Understanding the relation between protein structure and functionality is an important step in accomplishing these tasks.

Structure and Functionality

The ultimate goal of all researchers in this area has been to relate precisely the macroscopic manifestations of protein functionality in its utilization with its molecular properties. Great progress recently has been made in relating particular molecular structure to enzymatic activity. However, in the area of food functionality, the advance has been less rapid. The obvious difficulties are attributable to the mixed nature of food systems. The macroscopic manifestation in these products is usually the result of many interactions between different proteins and between these and nonprotein components. The multiplicity of the reactions under the conditions of food preparation or processing bears little resemblance to the clean-cut, precise reactions of pure proteins in solution. All extrapolations from the latter to the former have met with little success. In absence of this direct approach, we have resorted to various indirect approaches.

One method of bridging this gap has been the formulation of model systems. Usually these contain two or at most three components whose interactions are still observable and, it is hoped, closer to the multicomponent interactions found in actual utilization systems. Modified proteins, whose exact structure can be estimated, lend themselves to this approach.

Another avenue has been the study of protein molecular structure in the food mixture, before and after single process steps. This method endeavored to isolate the many food processing reactions into more investigable discreet steps.

A third approach, very commonly practiced, has been the study of specific properties in model systems of protein–nonprotein components of different but similar known proteins. The statistical correlation of the results to the behavior of the proteins in food systems throws some light on the effect of particular structural features on this behavior.

Any of these three methods of functionality investigation is open to various criticisms, especially on the basis of rigor. Therefore, in the final analysis we must resort to the simplest test of all scientific theories—predictability. If, after a careful study, the results of the elucidation of protein structural features predict its behavior in actual food utilization or processing, we conclude, a priori, that the structures are involved in the macroscopic manifestation. Otherwise, these tests, although very pretty and elaborate pieces of work, will only fatten the size of publications, with very little significance in relation to what we have attempted to do—understand the actual reasons for protein functionality.

Physicochemical Properties

There have been various attempts to classify functional properties and to relate them to specific physicochemical properties of the protein molecules (2, 3). The list compiled here is based on a recent one developed for specific soy proteins (1).

- Hydrophilic. These functional properties are influenced by the attraction of the proteins to water and its solutes: foaming, whipping, water binding, wetting, and stickiness.
- Hydrophilic–hydrophobic. These are mostly influenced by the amphoteric nature of proteins: emulsification, fat absorption, and fat holding.
- Intermolecular interactions. The properties governed by protein–protein reactions in the cold: film formation, whipping, viscosity, dough formation, and spreading.
- Thermal interaction. Properties strongly influenced by protein–protein interactions initiated or completely caused by thermal energy input: fiber formation, viscosity, gelation, film formation, elasticity, and texture.
- Specific effects. Some specific properties of the molecule: bleaching incompatibilities, stabilities, flavor, etc.

Also, it should be emphasized that each of these properties may, at times, have a general effect on the texture of the mixture which is a composite of the diverse specific functional properties of its many components. Thus, emulsification can, in numerous cases, govern both texture in general and juiciness in particular.

The 12 investigations compiled here deal mostly with functional investigations in the food area. Three chapters discuss soy proteins, relating the effects of heat, specific solvents, and enzyme activity on structure and functional properties. Single chapters are concerned with the proteins of wheat, milk, and yeast, respectively, as well as specific crosslinking effects and, enzyme–carrier interactions. Other chapters contrast general functional properties of different proteins.

This is, by no means, an all-inclusive volume on the subject. Rather, it is hoped that this volume will be a ground-breaking one to set some rules, to develop a consistent terminology, and to point the way to further research in this area. The results presented here no doubt will stimulate the authors and readers and, hopefully, will whet their appetite. Maybe in a year or two we will be able to have a longer, more specialized conference that will completely cover the subject.

We owe thanks to the following companies whose contributions aided us in financing some of the expenses of the participants: Miles Labs., the Pillsbury Co., the Archer Daniels Midland Co., and the Grain Processing Co.

Literature Cited

1. Pour-El, A., Measurement of Functional Properties of Soy Protein Products, "World Soybean Research," L. D. Hill, Ed., pp. 918–948, Interstate, 1976.
2. Wolf, W. J., Cowan, J. C., Soybean as a Food Source, *CRC Crit. Rev. Food Technol.* (1971) **2**, 81.
3. Kinsella, J. E., Functional Properties of Proteins in Foods: A Survey, *Crit. Rev. Food Sci. & Nutr.* (1976) **7**, 219.

PEACO   AKIVA POUR-EL
St. Paul, Minnesota
November 28, 1978

Oilseed Protein Properties Related to Functionality in Emulsions and Foams

JOHN P. CHERRY

Southern Regional Research Center, Science and Education Administration, Federal Research, U. S. Department of Agriculture, P. O. Box 19687, New Orleans, LA 70179

KAY H. McWATTERS and LARRY R. BEUCHAT

Department of Food Science, University of Georgia, College of Agriculture Experiment Station, Experiment, GA 30212

Proteins have molecular properties that contribute to functionality of food ingredients (1,2,3). Genetic and agronomic factors, as well as processing conditions, alter protein properties of seeds and, in turn, change the functional behavior of their products (2,4-10). Specific proteins contribute to the select functional behavior of a food; i.e., fractionating and reconstituting techniques have established that the proteins glutenin and gliadin are responsible for most of the functional, or breadmaking, properties of wheat flour. Various oxidizing and reducing reagents are used to alter sulfhydryl and disulfide bonds in these wheat proteins, producing varying bread quality. Data presented in this paper illustrate the importance of understanding the physical and chemical properties of proteins in seeds and seed products, so that their ultimate contributions as ingredients in foods can be maximized.

Efforts are under way to characterize the chemical properties of oilseed proteins and to relate these characteristics to the functional properties of seed products (9,11-16). Research deals with the chemistry of proteins affected by pH, suspension medium, heat denaturation, and enzyme degradation. Once constituents of seeds are identified and related to a specific functional property, efforts can be made to selectively modify or maintain these components to effectively utilize their beneficial contributions to food systems during processing, thereby expanding use of oilseeds and their by-products in foods.

Experimental Procedures

Soluble proteins in various aqueous fractions of glandless cottonseed flour (hexane-defatted) and peanut products (whole peanuts, and full-fat and hexane defatted meal and flour) were determined by methods of Cherry et al (10), and McWatters et

0-8412-0478-0/79/47-092-001$06.50/0

al. (13). Cherry et al. (10) described techniques for character-
izing the gel electrophoretic properties of soluble proteins in
the different preparations. Procedures for tests of functional
properties, including emulsion capacity and viscosity, and foam
capacity, stability, and viscosity, were those of McWatters and
Cherry (9). Hydrolysis of proteins with proteolytic enzymes
pepsin, bromelain, and trypsin was completed by the methods
of Beuchat et al. (8).

Results and Discussion

Functional Properties Related to pH

Protein Solubility. Proteins of glandless cottonseed, separated by centrifugation into soluble and insoluble fractions
from aqueous suspensions at pH values between 1.5 and 11.5,
produce solubility curves typical of vegetable proteins (Figure
1); these values, over the pH range of 1.5 to 11.5, were the
inverse of each other. For example, the quantity of soluble
protein declined in suspensions at pH values between 1.5 (acid-
dissociated proteins; 17) and 4.5 (isolectric point), and then
gradually increased as the pH was raised to neutrality; amounts
of storage proteins (11,18) increased as the pH was increased
to 11.5. These changes are further shown by the polyacrylamide
disc-gel electrophoretic patterns in Figure 2.

Emulsion Capacity and Viscosity. Glandless cottonseed flour
suspended in water can be emulsified to exhibit viscosities that
simulate those of salad dressing and mayonnaise (11). Varying
the pH of flour-in-water suspensions greatly influences emulsifying capacity and viscosity of cottonseed flour (Figure 1).
These changes in emulsifying capacity and viscosity can be
related to quantity and quality of protein present in the
soluble and insoluble portions obtained by centrifuging the
flour suspensions.
    Emulsion capacity of 10% suspension of flour, adjusted to
pH values between 1.5 and 11.5, ranged between 85 and 110 ml oil
per 25 ml suspension (Figure 1). The lowest and highest values
were noted at isoelectric pH (4.5) and 11.5, respectively; the
extent of solubility of proteins in the suspension correspond to
these values. Relative to pH, emulsion viscosity was highly
correlated with the percentage protein in the soluble fraction;
i.e., the viscosity declined between pH 1.5 and 4.5 (dissociated
polypeptides), and increased at first gradually, then rapidly,
as pH was adjusted toward 11.5 (water-extracted proteins are
soluble at neutral pH, and storage globulins at alkaline pH
values).

Foam Capacity, Stability, and Viscosity. It was more difficult to form foams of higher stability from suspensions of
glandless cottonseed flour as the pH was increased from 1.5 to

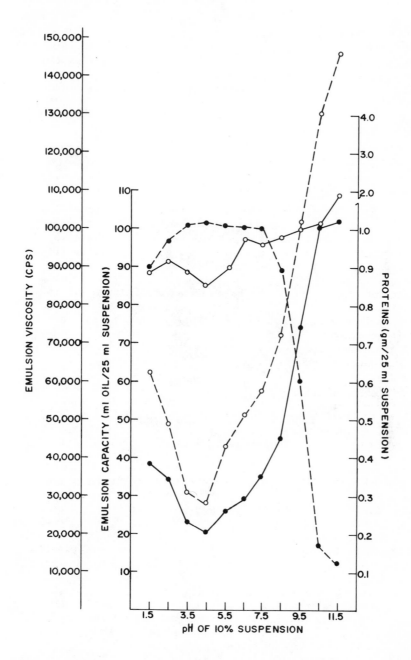

Figure 1.   Solubility and emulsifying properties of cottonseed proteins at various pH values. Emulsion: (○—○) capacity, (○– – –○) viscosity; protein fractions: (●—●) soluble, (●– – –●) insoluble.

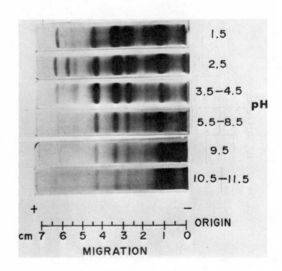

Figure 2. Gel electrophoretic properties of cottonseed proteins at various pH values

3.5 (Figure 3). Values were similar at pH 3.5 and 4.5. Between 4.5 and 6.5, foam capacity and stability declined to the lowest values noted in the experiment. As the pH was raised to 11.5, these functional properties of cottonseed flour increased to values similar to that of the pH 1.5; i.e., optimum foaming properties were noted when the dissociated proteins and major storage globulins were soluble.

Foam viscosity was optimum in the pH range of 3.5 to 5.5 (Figure 4). Between these values, most proteins of glandless cottonseed flour were in the insoluble form.

Functional Properties Related to Suspension Medium and pH

Protein Solubility. The effects of suspension medium and pH on selected functional properties of peanut protein were investigated (13) by suspending defatted peanut meal in water, 0.1 M NaCl, and 1.0 M NaCl (8% suspensions, w/v) and adjusting the pH to either 1.5, 4.0, 6.7, or 8.2; two-step sequential adjustments from 6.7 to 4.0 to 6.7 and from 6.7 to 4.0 to 8.2 were also included. Percentage of protein in soluble fractions from either water or salt suspensions was highest at pH 8.2, intermediate at 6.7, and lowest at 4.0 (Figure 5). Protein content was determined by macro-Kjeldahl analysis of nitrogen in the lyophilized fractions x 5.46. The isoelectric point of peanut seed proteins is between pH 3.0 and 5.0, depending on the type of suspension medium used to prepare these components (19). In extracts of water or in 0.1 M NaCl at pH 1.5, the percentage of proteins was similar to those at pH 4.0 and 6.7, or higher. The content of protein in soluble extracts of 1.0 M NaCl suspensions at pH 1.5 was less than that at pH 4.0. The changes in protein of insoluble preparations after each pH adjustment were the inverse of that of the soluble extracts (Figure 5).

The percentages of proteins in water-soluble extracts of suspensions of peanut meal adjusted from pH 6.7 to 4.0, then back to either 6.7 or 8.2, were less than those of the initial extract and the one-step pH change, respectively (Figure 5). Percentage of protein in soluble extracts at pH 6.7 or 8.2 was not altered greatly by the two-step pH adjustment.

Emulsion Capacity. The most viscous emulsions of peanut meal with mayonnaise consistencies were produced from suspensions at pH 1.5 in water and 0.1 M NaCl, and by the water suspension adjusted from pH 6.7 to 4.0 to 8.2 (Figure 6). At pH 4.0, poor emulsifying properties were noted for all suspensions. The two-step pH adjustment from 6.7 to 4.0 to 6.7 improved emulsification properties of only the water suspension over those of samples where the pH was not adjusted or was minor.

The high salt level (1.0 M NaCl) and poor protein solubility evidently affected emulsion formation at pH 1.5 (cf. Figures 5 and 6). At pH 6.7, levels of soluble proteins were variable in

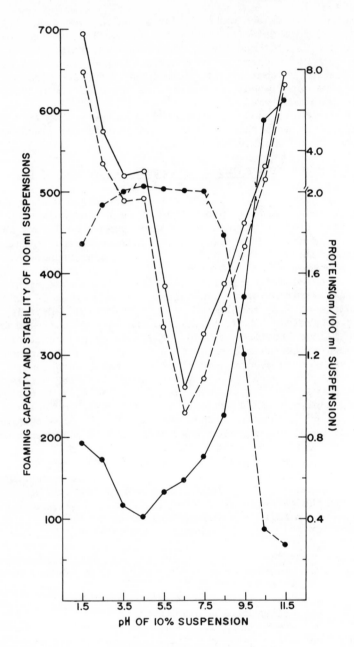

Figure 3. Solubility and foaming properties of cottonseed proteins at various pH
values. Foam: (○—○) capacity, (○– – –○) stability; protein fractions: (●—●)
soluble, (●– – –●) insoluble.

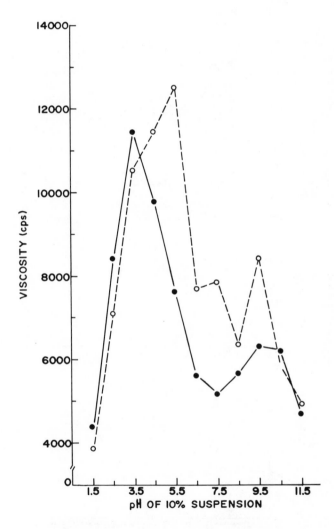

Figure 4.   *Foam viscosity and stability properties of cottonseed proteins at various pH values.* (●—●) *1 min,* (○– – –○) *60 min.*

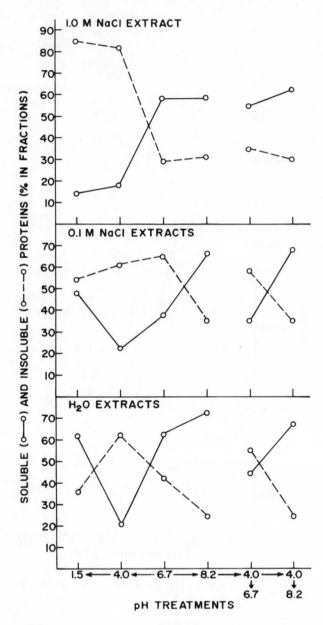

Figure 5. Effect of pH and suspension medium on the solubility properties of peanut proteins

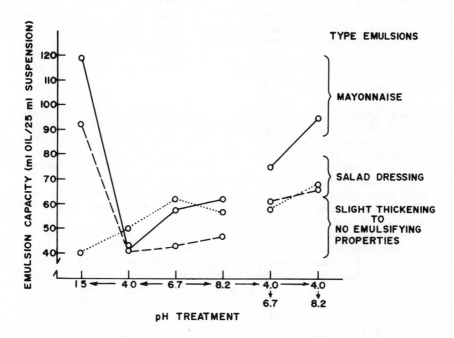

Figure 6.   Effect of pH and suspension medium on the emulsion capacity of pea-nut proteins. Suspension medium: (O—O) H₂O, (O– – –O) 0.1M NaCl, (O···O) 1.0M NaCl.

different suspensions. At pH 4.0, soluble protein levels were consistently low. None of these factors could be related to the ability of peanut meal to form emulsions (cf. Figures 5 and 6). Since the adjustment of pH from 6.7 to 4.0 to 6.7 or 8.2 for the water suspension improved emulsion capacity, and the others remained relatively unchanged, the salt may have interfered with this property. Percentages of protein in these suspensions remained relatively high. On the other hand, exposure of peanut meal components in suspension to pH 4.0, and then to pH 8.2, may have caused structural rearrangements in proteins soluble in water that did not occur in those of the salt solutions. A rearrangement in the structure of proteins in water would increase the number of binding sites to interact with oil and water. The slight decrease noted in solubility of proteins in water suspensions after the two-step pH change, compared to no pH adjustment or one-step adjustments suggested changes in the properties of these proteins (Figure 6).

Foam Capacity and Stability. Adjusting the pH of suspensions of peanut meal prior to whipping significantly affected foam formation and stability (Figure 7). The largest increase in foam volume occurred at pH 1.5; at this pH, increasing the salt concentration greatly reduced the percentage of protein in soluble fractions and depressed foam formation (cf. Figures 5 and 7). At pH 4.0, percentage of proteins in soluble fractions were low yet foam volume remained relatively high; little difference was noted between water and salt suspensions at this pH.

At pH 6.7 and 8.2, the smallest increases in foam volume due to whipping occurred (Figure 7). Percentage of protein in the soluble fraction of the pH 6.7 suspension was highly variable (Figure 5). Foam volume improved as salt concentration increased at this pH. Changing the pH from 6.7 to 4.0 to 6.7 improved the volume of foam formed over that of the unadjusted (pH 6.7) suspension, despite variation in percentage of protein in the soluble fractions.

Whether obtained by one- or two-step pH adjustment, at pH 8.2 the percentages of protein in soluble extracts were among the the highest observed in the study, yet increases in foam volume were not as great as those observed at pH 1.5 or 4.0.

Foam stability showed little variation among suspensions at pH 1.5 and 4.0, remaining between 35% and 45% (Figure 7). The largest variations in stability were noted among suspensions at pH 6.7 and 8.2, and after the two-step adjustment to pH 6.7. Increasing the amount of salt in the suspension at these pH values decreased foam stability. The suspensions adjusted from pH 6.7 to 4.0 to 8.2 exhibited similar foam capacity and stability, regardless of salt concentration.

McWatters and Cherry (9) characterized the emulsion and foam capacity, emulsion viscosity, protein solubility, and gel electrophoretic properties of water suspensions containing

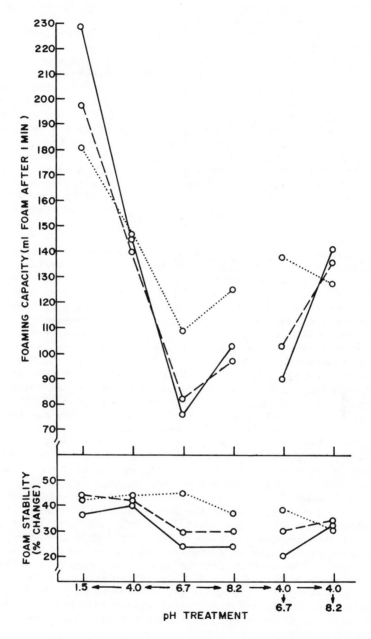

Figure 7. *Effect of pH and suspension medium on the foaming properties of peanut proteins. Suspension medium: (O—O) H_2O, (O---O) 0.1M NaCl, (O···O) 1.0M NaCl.*

defatted soybean, peanut, field pea, or pecan flour. The suspensions were evaluated at their natural pH, at pH 4.0, 8.2, and
after a two-step sequential adjustment from the natural pH to
4.0 to 8.2. Maximum functionality of suspensions was: (a)
soybean flour—very thick mayonnaise emulsions, and egg white-
type foams at natural pH; (b) peanut flour—semi-thick mayonnaise emulsions, and thick egg white-type foams after the two-
step adjustment to pH 8.2; (c) field pea flour—semi-thick mayonnaise emulsions and medium-thick foams at pH 8.2; and (d)
pecan flour—very thick mayonnaise emulsions after the two-step
adjustment to pH 8.2. Protein solubility seemed to be more
closely associated with improved viscosity of emulsions and
foams than with increased quantity of oil or air that could be
bound by flour suspensions. Data from gel electrophoretic
studies suggested that the major seed storage proteins were
important in functionality tests, although other seed constituents
such as carbohydrates may be equally involved.

    <u>Gel Electrophoresis of Proteins</u>. The proteins in various
suspension media (water, 0.1, 1.0 M NaCl) of peanut meal at
different pH values were characterized by electrophoretic techniques. All suspensions at pH 4.0 had few protein components
that could be distinguished on electrophoretic gels; more were
present as the salt level was increased in the suspensions
(Figure 8). No improvement in emulsion capacity occurred at
this pH, even though the number of protein components in the
soluble fraction increased in the presence of salt (cf. Figures
6 and 8). When the pH was lowered to 1.5, levels of nonarachin
proteins (region 2.5-4.5 cm) increased and the arachin components
(region 0.5-2.5 cm) became difficult to discern in the gels.
Investigators (<u>19</u>,<u>20</u>) have described these peanut protein components using gel electrophoretic techniques. Two components
in region 4.5-5.5 cm in gels of water and low salt extracts were
not present in those containing 1.0 M NaCl. Only the former two
preparations formed emulsions resembling mayonnaise; they also
produced the greatest increases in foam volume. At pH 6.7, 8.2,
and the two-step pH adjustments, water and low salt mixtures had
soluble proteins with similar electrophoretic properties. These
gels showed that suspensions with 1.0 M NaCl had lower amounts
of nonarachin proteins in region 2.5-5.5 cm. This characteristic
could be related to the poor emulsifying properties of these
preparations. The arachin components in these high salt solutions
were diffuse and difficult to discern in the gels. No specific
protein properties detectable by gel electrophoresis could be
related to the unique ability of the water suspension subjected
to the two-step pH adjustment of 6.7 to 4.0 to 8.2 to form a
mayonnaise emulsion.
    Dough quality and strength in breakmaking depends on the
gluten proteins in wheat (<u>2</u>). Hyder <u>et al</u>. (<u>21</u>) demonstrated by
starch gel electrophoresis that the isoelectric protein fraction

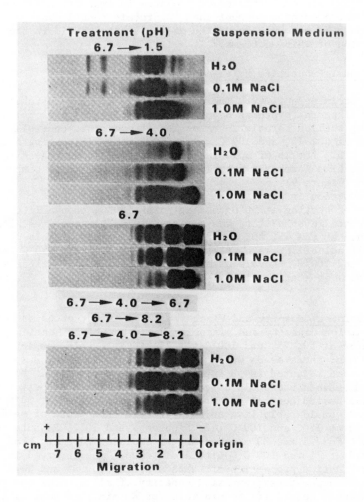

Figure 8.   Gel electrophoretic properties of peanut proteins as influenced by pH and suspension medium

of soybean flour and a synthetic glycolipid interacted during
dough mixing with the gluten fraction of wheat flour. This interaction may be the basic mechanism contributing to loaf volume.
In meat emulsions, a specific soluble fraction of proteins was
shown to function as the key emulsifier (22). However, Lin et al.
(15) showed that sunflower meal was superior to soybean and sunflower protein concentrates or isolates in emulsification capacity. These authors suggest that nonprotein constituents of seeds
may contribute to the formation of emulsions and aid in the
formation of whipped foams.

Functional Properties Related to Heat

   Protein Solubility. The effects of moist heat on functionality of peanut proteins have been investigated (10,12,14) by
heating shelled kernels in water in a temperature-controlled
retort at 50, 75, and 100°C for 15-min intervals ranging from
15-90 min. Levels of soluble protein generally decreased as
heating time increased from 15 to 90 min at all three temperatures (Figure 9).
   Reduction in protein solubility was somewhat greater after
15 min of heating at 50°C than at 75 or 100°C, though little
difference in solubility was observed among temperatures after
30 min. At the 45, 60, and 75 min heating times, levels of
protein solubility were greater in peanuts heated at 50°C than
at 75 and 100°C. Levels of soluble protein were drastically
reduced when heating time at 100°C was extended to 60, 75, and 90
min.

   Emulsion Capacity and Viscosity. The effect of moist heat
on emulsion capacity and viscosity was measured by heating peanuts
in water at 50, 75, and 100°C for 15-min intervals ranging from
15-90 min; these seeds were than ground into a paste, suspended in
water, and adjusted in pH from 6.7 to 4.0 to 8.2 (14). Moist-
heated peanuts had emulsion capacities similar to or better than
the non-heated control (Figure 10). Peanuts heated at 50°C for 15
min had considerably less soluble protein than those of seeds
treated at 75°C and 100°C (cf. Figures 9 and 10). Peanuts heated
at 50°C for 15 and 30 min had higher emulsion capacity than those
treated at 75 and 100°C. Little difference was observed in
emulsification properties of samples heated for 45 and 60 min at
various temperatures; emulsion capacity of all samples was greater
for peanuts heated for 60 min than for 45 min. As heating time
was extended to 90 min, emulsion capacity of suspensions made with
peanuts heated at 50°C and 100°C improved, whereas a reverse trend
was observed for samples heated at 75°C. In the former two treatments, protein solubility continually declined, but in the latter
case, an increase in protein solubility was noted at 90 min.
   Although all emulsions in this study were thick and peaked,
viscosities decreased as the treatment temperature was increased

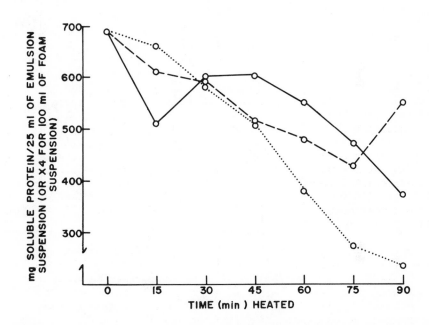

Figure 9. Quantity of soluble protein in peanuts moist heated for various time intervals. Suspension made with peanuts moist heated at: (O—O) 50°C, (O----O) 75°C, (O···O) 100°C.

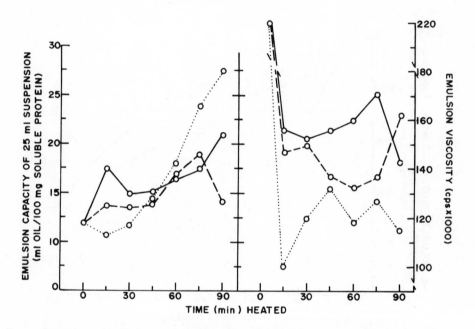

Figure 10.   Emulsifying properties of suspensions made from peanuts moist heated for various time intervals. Suspension made with peanuts moist heated at: (○—○) 50°C, (○ – – –○) 75°C, (○ · · · ○) 100°C.

(Figure 10). All heat-treated samples were less viscous than the unheated control. Viscosity remained similar for samples heated at 50°C for 15 to 45 min, increased between 60 and 75 min, then declined at 90 min. In general, protein solubility declined in all of these suspensions. At 75°C, the viscosities declined for peanuts heated for 15 to 60 min, then increased if heated for 90 min. Similar observations were noted for protein solubility in these suspensions. Emulsion viscosities of peanuts heated at 100°C increased between 15 and 45 min, then remained fairly constant between 45 and 90 min; protein solubility of these suspensions declined continually between 15 and 90 min.

Foam Capacity and Stability. The effect of moist heat on foam capacity and stability were measured by moist heating peanuts at 50, 75, and 100°C for 15-min intervals ranging from 15-90 min; these seeds were then ground into a paste, suspended in water, and adjusted in pH from 6.7 to 4.0 to 8.2 (14). Foam capacity of peanuts heated at 50°C increased slightly during the 15 to 90 min heating period (Figure 11), but showed no major changes during this same heating period at 75°C. At 100°C, samples heated for 15 to 90 min produced the largest increase in foam capacity. In most cases, foam capacity was inversely related to protein solubility; i.e., where protein solubility decreased, foam capacity increased, and vice versa (cf. Figures 9 and 11).

Foams prepared from peanuts heated at 75 and 100°C varied, but were more stable than those of seeds heated at 50°C for different heating times to 90 min (Figure 11).

Gel Electrophoretic Properties. Gel electrophoretic patterns of soluble proteins revealed that heating peanuts at 50 and 75°C up to 90 min caused little detectable change in the soluble proteins (Figure 12). Heating peanuts at 100°C for various time intervals to 90 min altered the structural components of non-arachin proteins in region 1.5-3.5 cm of gels. The arachin fraction (region 0.5-1.5 cm), which is the major storage globulin of peanuts, was unchanged on electrophoretic gels of samples heated at 100°C for 15 to 90 min. During this same heating period at 100°C both emulsion and foam capacity increased greatly. Emulsion viscosity of these suspensions was low, however, foam stability was among the best. The arachin was gradually altered to other forms (polypeptide subunits and/or aggregates; 10) by extended heating from 105 to 180 min (Figure 12). Similar changes occurred in the arachin fraction of peanuts heated at 120°C for 15 min, being accelerated by treatment at this relatively high temperature (10). Heat treatments of 15 min at 120°C and more than 105 min at 100°C produced pourable, suspension-type products rather than thick emulsions. These data suggest that qualitative changes in nonarachin protein content due to moist heating of peanuts at 100°C for time intervals of 15 to 90 min

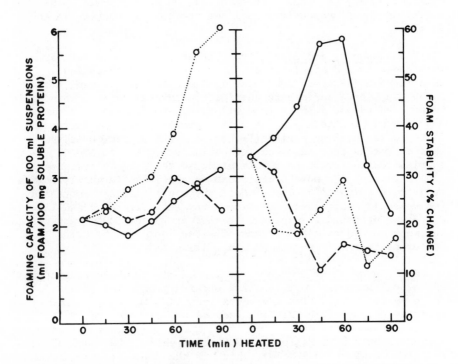

Figure 11. Foaming properties of suspensions made from peanuts moist heated for various time intervals. Suspension made with peanuts moist heated at: (O—O) *50°C,* (O– – –O) *75°C,* (O···O) *100°C.*

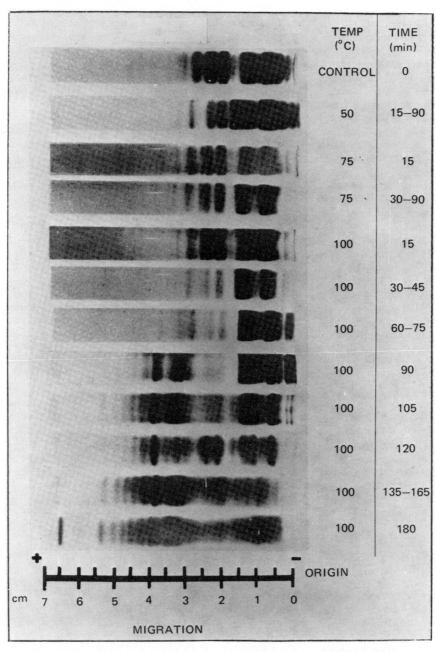

Figure 12. *Gel electrophoretic properties of soluble proteins in peanuts moist heated for various time intervals (10)*

may be related to the improved ability of pastes prepared from
these samples to form emulsions and foams. On the other hand, the
combinations of arachin and the denatured forms of nonarachin
proteins may contribute to emulsion and foam formation. At 50°C
and 75°C, these changes may be occurring but are not large enough
to be detectable by gel electrophoretic techniques; changes in
protein solubility of these preparations suggest that protein
structural changes are occurring.

     Functional properties of proteins derived from various sources
have been shown to be affected by heat. For example, heating
protein dispersions of soybeans (23) and Great Northern beans (24)
improved their ability to form foams, whereas with sunflower (25)
such treatment had an adverse effect. Foaming capacity of peanut
proteins could probably be improved by removal of certain components that interfere with foam formation. Eldridge et al. (26)
found that washing acid-precipitated soybean protein with aqueous
alcohol removed a phospholipid-like material and improved foam
formation and stability. In foaming studies with sunflower meal
(15), nonprotein constituents apparently aided the formation of
whipped foams. In meat emulsions, a specific soluble fraction
of proteins was shown to function as the key emulsifier (22).
However, the moist heat study described here (10,14) was conducted
with whole peanuts rather than protein concentrates or isolates,
thus the role of carbohydrate, lipids, and/or complexes of these
components that are formed during moist heating and subsequent
blending of peanut paste-water suspensions with oil is unknown.
Lin et al. (15) showed that sunflower meal was superior to soybean
and sunflower concentrates or isolates in emulsion capacity.
McWatters and Cherry (9) compared select functional properties of
defatted soybean, peanut, field pea, and pecan flours and showed
that major seed storage proteins were important in emulsifying and
foaming properties. Protein solubility was related to the quality
of the emulsions and foams. Behavioral characteristics contributed by nonprotein components that occur naturally in the seeds,
especially carbohydrates, were implicated.

Functional Properties Related to Enzyme Hydrolysis of Proteins

     Protein Solubility. The effects of proteolysis on function-
ality of peanut flour has been investigated (8,27-29). Protein
solubilities of nontreated peanut flour and trypsin-treated
flour (0 min) at pH 7.6 were similar, but were lower for those
with bromelain at pH 4.5 and pepsin at pH 2.0 (Figure 13); the
pH values of the latter two enzyme treatments were readjusted
back to 7.6 for these analyses. Lowering the pH of the pepsin-
treated control to 2.0 and that of the bromelain control to 4.5,
followed by heating to stop enzyme activity, evidently altered
peanut proteins irreversibly to less soluble forms. Adjusting
the trypsin-treated control to pH 7.6 during the heating procedure did not require passing the proteins through their

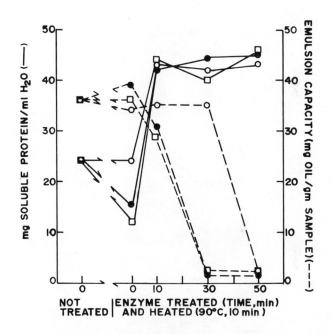

Figure 13. Solubility and emulsifying properties of proteins in peanut flour treated with proteolytic enzymes. (◪) Not treated, pH 6.9; (○) trypsin, pH 7.6; (□) bromelain, pH 4.5; (●) pepsin, pH 2.0.

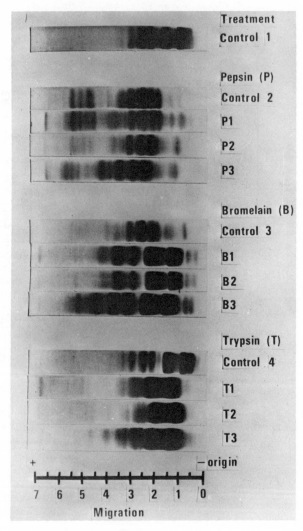

Figure 14.   Gel electrophoretic properties of proteins in peanut flour treated with proteolytic enzymes (8)

isolectric points as was done with the other two enzyme treatments. This procedure did not significantly change the solubility of peanut proteins.

Enzymatic hydrolysis increased the soluble protein, compared
to the respective controls (Figure 13). Approximate threefold
increases were noted after pepsin digestion, fourfold after
bromelain, and twofold after trypsin. These levels of protein
solubility were evident after only 10 min of enzyme treatment.
Further digestion times did not significantly increase the amount
of soluble protein.

Gel Electrophoresis. Electrophoretic patterns of proteins
in untreated and enzymatically hydrolyzed peanut flour showed
that the proteins of pepsin and bromelain controls were markedly
different from those of nontreated flour (Figure 14). Heating
these two controls under acidic conditions not only decreased
protein solubility, but also changed the proteins qualitatively.
No major arachin components were detected in the 0.5-1.5 cm
region of gels of the pepsin and bromelain controls. Two minor
bands were detected in region 0.5-1.5 cm, and a number of major
components were clearly shown in regions 2.0-4.0 and 4.5-5.5 cm
of gels of the pepsin control. The gel pattern of the bromelain
control showed three minor components in region 0.5-1.5 cm and a
number of dark-staining bands in region 2.0-4.0 cm.

Digestion of peanut flour with pepsin and bromelain resulted
in considerable changes in the lower half of the gels (region
4.5-7.0 cm). The changes in the protein patterns of peanut flour
treated with these enzymes, especially pepsin, became most pre-
valent by 50 min. The gel pattern of the bromelain-digested flour
did not resemble its control or the nontreated flour. Six major
bands were distinctly shown in region 1.0-5.0 cm, and four minor
components were present in region 0.5-6.8 cm that were not clearly
distinguished in the controls. The dark-staining components in
region 1.0-2.5 cm had mobilities similar to the major diffuse
bands detected in the trypsin-treated samples, and may be altered
forms of arachin, the major storage globulin in peanuts (20).

The gel pattern of the trypsin control was very similar to
that of the nontreated peanut flour. This observation supports an
earlier suggestion that adjusting the pH of the trypsin control to
7.6, followed by heating, did not alter the proteins enough to
substantially affect their solubility. However, qualitative
changes were noted in the soluble protein fraction of trypsin-
treated flour. Arachin appears in the 0.5-1.5 cm region of the
electrophoretic pattern of nontreated flour, but is not clearly
shown in trypsin-treated preparations. Instead, two diffuse
major bands were detected in region 1.0-2.5 cm, and a number
of smaller components were present in the lower halves of the
gels. These major bands may result from arachin, which was
partially digested by trypsin to yield forms having increased
electrophoretic mobilities.

Emulsion Capacity. Enzymatic digestion of proteins beyond
10 min, except the trypsin-treated sample for 30 min, destroyed
emulsifying capacity of the flour (Figure 13). Apparently,
hydrolysis substantially altered protein surface activity
strengths and the ability of the protein to stabilize oil-in-
water emulsions. This assumption agrees with earlier work
showing decreased emulsion capacity of peanut flour fermented
with fungi (27).

Conclusions

Understanding protein changes that occur during processing
of oilseeds as food products is prerequisite to accurately
predicting the functional behavior of these nutritious constituents in food systems. Genetic factors and growing location, as
well as harvesting, curing, and storage practices, add to the
complexity of chemical properties of the proteins stored in
oilseeds. Varying levels of protein, salt, pH, and heat, in
addition to molecular interactions among seed storage constituents, affect certain physicochemical properties in suspension
media used in food formulations. Proteins denature to various
dissociated or aggregated forms and are detectable by gel electrophoretic techniques as changes in size, conformation, and charge.
The level or proportion of certain soluble proteins is a measure
of the availability of these components for functional expression.

Literature Cited

1. Finney, K. F., In "Postharvest Biology and Biotechnology,"
   Hultin, H. O., and Milner, M., eds., Food and Nutrition
   Press, Inc., Westport, Conn. (1978).

2. Huebner, F. R., Bietz, J. A., and Wall, J. S., In "Protein
   Crosslinking - Biochemical and Molecular Aspects,"
   Friedman, M., ed., 67, Plenum Press, New York, N. Y.
   (1977).

3. Ryan, D. S., Adv. Chem. Ser. (1977) 160:67.

4. Richardson, T., Adv. Chem. Ser. (1977) 160:185.

5. Goforth, D. R., Finney, K. F., Hoseney, R. C., and
   Shogen, M. D., Cereal Chem. (1977) 54:1249.

6. Patil, S. K., Finney, K. F., Shogren, M. D., and Tsen, C. C.,
   Cereal Chem. (1976) 53:347.

7. Beuchat, L. R., J. Agric. Food Chem. (1977) 25:258.

8. Beuchat, L. R., Cherry, J. P., and Quinn, M. R., J. Agric. Food Chem. (1975) 23:616.

9. McWatters, K. H., and Cherry, J. P., J. Food Sci. (1977) 42:1444.

10. Cherry, J. P., McWatters, K. H., and Holmes, M. R., J. Food Sci. (1975) 40:1199.

11. Cherry, J. P., Berardi, L. C., Zarins, Z. M., Wadsworth, J. I., and Vinnett, C. H., In "Improvement of Protein Nutritive Quality of Foods and Feeds," Friedman, M., ed., Plenum Press, New York, N. Y. (1978).

12. Cherry, J. P., and McWatters, K. H., J. Food Sci. (1975) 40:1257.

13. McWatters, K. H., Cherry, J. P., and Holmes, M. R., J. Agric. Food Chem. (1976) 24:517.

14. McWatters, K. H., and Cherry, J. P., J. Food Sci. (1975) 40:1205.

15. Lin, M. J. Y., Humbert, E. S., and Sosulski, F. W., J. Food Sci. (1974) 39:368.

16. Wolf, W. J., and Cowan, J. C., Crit. Rev. Food Technol. (1971) 2:81.

17. Berardi, L. C., and Cherry, J. P., Cereal Chem. (1978) In press.

18. Berardi, L. C., Martinez, W. H., and Fernandez, C. J., Food Technol. (Chicago) (1969) 23(10):75.

19. Basha, S. M. M., and Cherry, J. P., J. Agric. Food Chem. (1976) 24:359.

20. Cherry, J. P., Dechary, J. M., and Ory, R. L., J. Agric. Food Chem. (1973) 21:652.

21. Hyder, M. A., Hoseney, R. C., Finney, K. F., and Shogren, M. D., Cereal Chem. (1974) 51:666.

22. Saffle, R. L, Adv. Food Res. (1968) 16:105.

23. Eldridge, A. C., Hall, P. K., and Wolf, W. J., Food Technol. (Chicago) (1963) 17(12):120.

24. Satterlee, L. D., Bembers, M., and Kendrick, J. G., J. Food Sci. (1975) 40:81.

25. Huffman, V. L., Lee, C. K., and Burns, E. E., J. Food Sci.
     (1975) 40:70.

26. Eldridge, A. C., Wolf, W. J., Nash, A. M., and Smith, A. K.,
     J. Agric. Food Chem. (1963) 11:223.

27. Quinn, M. R., and Beuchat, L. R., J. Food Sci. (1975)
     40:475.

28. Beuchat, L. R., Lebensm. Wiss. Technol. (1977) 10:78.

29. Beuchat, L. R., Cereal Chem. (1977) 54:405.

RECEIVED October 16, 1978.

Effect of Conformation and Structure Changes Induced by Solvent and Limited Enzyme Modification on the Functionality of Soy Proteins

B. A. LEWIS and J. H. CHEN

Division of Nutritional Sciences, Cornell University, Ithaca, NY 14853

Improvement in functionality of the major soy globulins can be obtained by structural modification of the proteins. Frequently this is done by chemical modification of the lysine epsilon-amino groups or by alkaline or heat treatment. We have described a procedure for preparation of a superior soy product without alteration of the amino acids (1).

Limited digestion of the globular soy proteins (Harosoy variety) with the enzyme rennin affords a modified protein preparation which retains a high molecular weight. The enzyme-modified protein is precipitated and washed with alcohol and subsequently heat-modified. A flavorless product results which is easily dispersed in water and shows excellent functional characteristics.

Whipping quality, measured by foam volume and stability, is superior in comparison with the native proteins. The product has excellent water-binding properties and good heat gelation characteristics. The limited rennin proteolysis of the soy protein is a key point in the functionality of the protein product since this modification confers improved solubility in water. In addition, for good whippability and gelation, it is important that the proteins have a high molecular weight and this has been demonstrated by viscosity data, gel filtration, and polyacrylamide gel electrophoresis. As would be anticipated from the limited proteolysis, a good yield of protein is recovered from the enzyme treatment.

It is of interest to obtain quantitative information concerning the role of chemical structure and conformation on the functional properties of proteins. Such information is useful in determining effective methods for altering functionality or in plant breeding to build in desirable traits or increase the yields of those proteins having more desirable properties.

This paper describes the characteristics of the major soy globulins, the primary and higher structure of the 7S globulin, and structural characteristics of the 7S protein modified by rennin digestion and subsequent alcohol and heat denaturation.

Soy Globulins

Soy beans contain several globular proteins which can be separated from the whey proteins by isoelectric precipitation at pH 4.5-4.9. About 90% of the extracted protein is recovered in this isolate. Of the globular proteins, the 7S and 11S proteins predominate although the actual amount of each varies with soy variety. The isolation and properties of these proteins has been the subject of recent reviews (2, 3). Some of the undesirable features of soy proteins are illustrated by the following observations.

The water-extractability of both the 7S and 11S proteins decreases with aging of the soy meal (4). During isoelectric precipitation 15-30% of the globular fraction is denatured and fails to redissolve in neutral buffer (4, 5). As the pH is lowered to 2-3 the 11S component becomes increasingly sensitive to irreversible denaturation. This acid-sensitive fraction not only limits the solubility of such isoelectric protein isolates, and therefore their functional uses, but this fraction may also be responsible for retention of undesirable flavors (5).

Soy globulins also have a tendency to undergo a reversible dimerization with changes in ionic strength, the 7S globulin converting to a 9S dimer at low ionic strength (2).

Characteristics such as these have complicated the isolation, purification, and structural determination of the individual globular proteins. Nevertheless, some structural and conformational information is available for the 7S and 11S globulins.

Primary Structure of the 7S Protein

The 7S globulin is a glycoprotein containing 3.8-5.4% carbohydrate. Mannose and glucosamine appear to be the only sugars present and they are linked through glucosamine to asparagine (6). The amino acid composition has been reported for various preparations of 7S protein (7, 8, 9, 10) but the results vary significantly. This may represent the level of purification of the protein or varietal differences. Nevertheless, all studies show that the sulfur amino acid content is very low and sulfhydryl groups are probably not present. Two disulfide linkages may be present however.

The estimated molecular weight is in the range 150,000-185,000 (8, 10) for the monomeric 7S and 370,000 for the dimeric 9S form (10). As many as eight N-terminal amino acids have been reported (8) although more recently only valine and leucine have been detected at the N-terminal (10). Beyond these observations, the primary structure has not been determined.

Secondary and Tertiary Structure

The secondary and tertiary structure of a partially purified 7S globulin was examined by Fukushima (7) based on optical rotatory dispersion, infrared and ultraviolet difference spectra. Antiparallel β-structure (35%) and random coil (60%) predominated with only 5% helical structure present. The contribution of the three structures was calculated from molecular ellipticity values obtained by circular dichroism (11) and from the Moffitt parameters in ORD (11, 12). Between 210 and 250 nm, the experimental CD curve for the 7S protein was similar to the CD curve computed from ORD Moffitt parameters with the major dissimilarity occurring at 208-213 nm.

Our studies on 7S globulin isolated under conditions which avoided the acidic denaturation conditions created by isoelectric precipitation showed about 5% helix, 60% β-structure and 35% random coil. It is quite possible that the isolation conditions have determined the secondary order of the isolated protein.

Fukushima (7) observed that 40% of the peptide hydrogens were not exchanged with deuterium and concluded that the molecules are compactly folded even in the random areas. The intrinsic viscosity of 0.0638 dl per g reported by Koshiyama is also consistent with a relatively compact structure (8, 9).

In the 7S globulin, tryptophan appears to be located on the surface while tyrosine is buried in the internal hydrophobic region (7, 8). In the UV spectrum of the 7S globulin (pH range 2.50-7.93), four shoulders appearing between 250-280 nm are attributed to phenylalanine absorption (13). A slight shift of these shoulders as well as the 250 nm minimum to longer wave lengths was observed at pH 10. Ionization of the tyrosine hydroxyl groups occurs in 0.1 N sodium hydroxide resulting in a red shift of the total spectrum.

A similar secondary structure is predicted for the 11S globulin. However, differences appear in the tertiary structure as shown by the CD spectrum in the 240-320 nm region. Both tyrosine and tryptophan appear to be buried in the hydrophobic interior of the 11S globulin. CD spectra for both proteins in 0.1 N sodium hydroxide also reveal significant differences in tertiary structure (11). Tyrosine residues in the 7S protein appear to ionize much more readily than in the 11S protein as determined by the magnitude of the positive peak at 253 nm.

Preliminary evidence from enzymatic digestion of the native 7S and 11S proteins suggests that the 7S protein may have a more hydrophilic surface than the 11S protein. Thus, trypsin digests the 7S fraction more rapidly than the 11S at pH 5.8-6.7, whereas rennin degrades the 11S more rapidly than the 7S. The epsilon-amino groups of lysine may also be more available on the surface of the 7S resulting in easier succinylation.

Quaternary Structure

Thanh and Shibasaki (10) proposed a trimeric structure for the 7S and a hexameric structure for the 9S dimer. Urea/sodium dodecyl sulfate polyacrylamide gel electrophoresis resolves the 7S globulin into six isomeric forms which are made up of three types of subunits (α, α' and β) in varying proportions (10, 14, 15). The composition of the six isomeric proteins has been designated as follows: B_1, $\alpha'\beta_2$; B_2, $\alpha\beta_2$; B_3, $\alpha\alpha'\beta$; B_4, $\alpha_2\beta$; B_5, $\alpha_2\alpha'$ and B_6, α_3. Molecular weights of 42,000 and 57,000 have been reported for the subunits. The six isomeric forms showed ultraviolet absorption spectra similar to that reported originally by Koshiyama (13) with maxima at 278 nm and at 232-234 and a minimum at 250 nm with shoulders at 253, 258, 265 and 269 (10).

Structural Changes in the Modified 7S Protein

As already described the enzyme-modified 7S protein retains a high molecular weight; like the native protein it is excluded by Bio-Gel P-150 which has an exclusion limit of 150,000 daltons. The specificity of rennin is such that it is easy to control and limit the extent of digestion. When the enzyme action is monitored by ultraviolet absorption it is apparent that the rennin action is quite different from that obtained with an enzyme such as trypsin (Figure 1). Thus the UV difference spectrum for the rennin-modified protein shows an initial unfolding of the 7S protein chains as indicated by a negative peak at 236 nm. As rennin action continued this negative peak was replaced by a positive peak at about 237 nm characteristic of an ordered secondary structure.

In contrast trypsin digestion continued to cause an unfolding of the protein chains disrupting the secondary and the tertiary structure and exposing the tyrosine and tryptophan residues which had been buried in the hydrophobic folds of the native protein (Figure 1). The spectrum displayed by the trypsin-digested protein is similar to the 7M urea-denatured protein (Figure 2) with a strong minimum at 233 nm and weak minima at 278 and 285. The strong minimum at 236 in trypsin-modified 7S protein may reflect destruction of the primary as well as the secondary structure.

It is well established that monohydric alcohols perturb the hydrophobic interior of the native globular proteins. The effectiveness of the water-miscible alcohols as protein denaturants increases with their increasing chain length and hydrophobicity (16, 17). Thus the denaturing action is due to the effect on hydrophobic bonding in the protein (16, 17). The unfolding and disorganization of the interior hydrophobic region of the globular protein is followed by a refolding of the polypeptide chain with a net increase in helical association or other intra

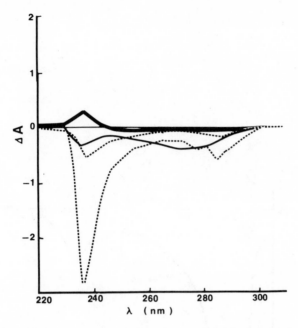

Figure 1.   *Difference spectra for soy 7S protein modified by enzymatic hydrolysis (– – –) by trypsin at pH 7.1 and (——) by rennin at pH 5.8*

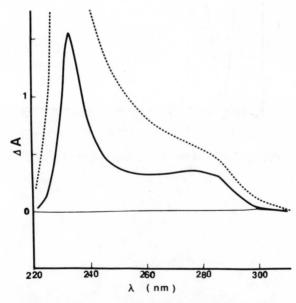

Figure 2.   *Difference spectrum for 7M urea-denatured soy 7S protein. Five minutes in (——) 20% ethanol and (– – –) in 35% ethanol.*

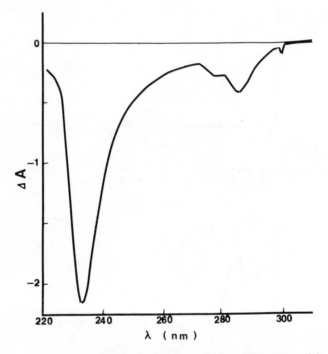

Figure 3. Difference spectra for alcohol-denatured soy 7S protein. Fifteen minutes at room temperature.

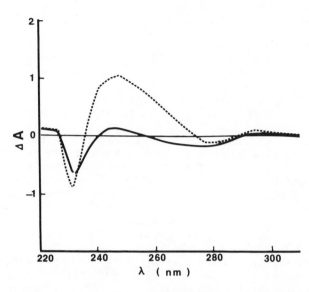

Figure 4.   Difference spectra for heat-denatured and rennin/alcohol-modified soy 7S and 11S protein fraction. (– – –) Heat denatured, (——) enzyme modified.

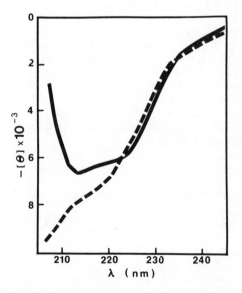

Figure 5.   Circular dichroism curves for soy 7S protein and modified 7S protein.
(——) Native, (– – –) modified.

and interchain associations (16).

The disruption of the hydrophobic interior and exposure of the tyrosine residues in the enzyme and alcohol-modified protein is revealed by the appearance of the positive peaks in the region of 276-278 nm and 284 nm in the UV difference spectrum (Figure 3). The subsequent refolding to a helical or other ordered structure is shown by the strong positive peak at 233 nm.

The effect of heat denaturation on the native mixed soy 7S and 11S globulins and on the enzyme-modified soy 7S and 11S proteins is shown by the UV difference spectra (Figure 4). The strong negative peak at 232-233 nm indicates rupture of the secondary structure which has occurred in both the native and enzyme-modified protein. The enzyme-modified protein may retain more order and be somewhat more stable to heat denaturation than the native protein. Loss of secondary structure is also apparent in the circular dichroism spectra (Figure 5) with the change in character of the curve below 220 nm. The spectrum resembles that reported by Koshiyama and Fukushima (11) for 7S protein in 0.25% SDS/0.01M tris buffer.

For the native protein the CD spectrum shows a large minimum at 216-217 nm and a shoulder at 222 (Figure 5). This is similar to the spectrum reported by Arai et al. (18). Koshiyama and Fukushima (11) reported a minimum near 210 nm with a shoulder at 222 nm.

Thus, the experimental evidence shows that the native 7S globular protein of soy beans is a high molecular weight compactly folded protein possessing primarily random coil and β-structure. Rennin action causes a physical refolding of the molecular chains but has relatively slight effect on the molecular weight. The hydrophobic domain is perturbed by the subsequent alcohol treatment and the protein chains assume a more helical structure. As would be anticipated the secondary structure in both the native and modified protein is disorganized by the heat treatment. Although the structural changes effected by rennin are slight, the modified protein has significantly improved functional properties.

Literature Cited

1. Chen, J. H. and Lewis, B. A., Abstracts, Inst. Food Tech. Meeting, Philadelphia (1977).
2. Wolf, W. J. in "Soybeans: Chemistry and Technology, Vol. 1, Proteins," eds. A. K. Smith and S. J. Circle, 93-143, AVI Publ. Co., Westport, CT. 1972.
3. Wolf, W. J., J. Am. Oil Chem. Soc. (1977) 54, 112A.
4. Nash, A. M., Kwolek, W. F. and Wolf, W. J., Cereal Chem. (1971) 48, 360.
5. Anderson, R. L., Cereal Chem. (1974) 51, 707.
6. Yamauchi, F., Kawase, M., Kanbe, M. and Shibazaki, K., Agric. Biol. Chem. (1975) 39, 873.

7. Fukushima, D., Cereal Chem. (1968) 45, 203.
8. Koshiyama, I., Cereal Chem. (1968) 45, 394.
9. Eldridge, A. C. and Wolf, W. J., Cereal Chem. (1969) 46, 470.
10. Thanh, V. H. and Shibasaki, K., J. Agric. Food Chem. (1978) 26, 692.
11. Koshiyama, I. and Fukushima, D., Cereal Chem. (1973) 50, 114.
12. Koshiyama, I., Agric. Biol. Chem. (Tokyo) (1970) 34, 1815.
13. Koshiyama, I., Cereal Chem. (1968) 45, 405.
14. Thanh, V. H., Shibasaki, K., Biochim. Biophys. Acta (1976) 439, 326.
15. Thanh, V. H., Shibasaki, K., Biochim. Biophys. Acta (1977) 490, 370.
16. Herskovits, T., Gadegbeku, B. and Jaillet, H., J. Biol. Chem. (1970) 245, 2588.
17. Fukushima, D., Cereal Chem. (1969) 46, 156.
18. Arai, S., Noguchi, M., Yamashita, M., Kato, H. and Fujimaki,M. Agric. Biol. Chem. (Tokyo) (1970) 34, 1338.

RECEIVED September 27, 1978.

Chemical Modification for Improving Functional Properties of Plant and Yeast Proteins

J. E. KINSELLA and K. J. SHETTY

Department of Food Science, Cornell University, Ithaca, NY 14853

People in many regions of the world suffer from protein-calorie malnutrition. If the world population continues to expand toward seven billion by the 21st century the need for more protein will become accentuated. As real demand for protein increases with a burgeoning population and as the emphasis changes from conventional agriculture (because of limitations of land and energy) to more direct consumption of plant and microbial proteins the necessity for new processes and new products will increase.

There are large resources of potential food proteins (oilseed, yeast, leaf) which are presently unexploited. With the application of innovative scientific and technological methods these can become significant sources of food protein. In developing ingredient protein from plant sources, research emphasis must include studies to determine the physicochemical or functional properties of these proteins.

Functional properties determine the overall behavior or performance of proteins in foods during manufacturing, processing, storage, and consumption ($\underline{1}$). They reflect those properties of the protein that are influenced by its composition, its conformation, and its interactions with other food components as affected by the immediate environment (Table I).

Typical functional properties include emulsification, which is important in sausage-type processed meats and coffee whiteners; hydration and water binding, which are critical in doughs and meat products; viscosity, important for beverages, e.g. liquid instant breakfasts; gelation, required in marshmallows and cold meat products; foaming/whipping, vital in whipped toppings; cohesion, which is important in manufacture of textured products; and color control, which is important in several products, particularly breads. These criteria were often overlooked in discussing new protein sources where quantity of protein and its biological value were the only considerations. The successful supplementation of existing foods, the replacement or simulation of traditional

0-8412-0478-0/79/47-092-037$06.75/0

proteinaceous foods and the fabrication of new foods sometimes
depends on the availability of proteins with these critical functional characteristics.

Table I. Functional Properties of Proteins Important in Food
          Applications

General Property	Functional Terms
Hydration	Solubility, dispersibility, wettability, water absorption, water holding capacity, swelling, thickening
Surface Activity	Emulsification, foaming (aeration whipping), protein/lipid film formation, lipid flavor, pigment binding
Textural Rheological	Elasticity, viscosity, grittiness, cohesiveness, chewiness, adhesion, aggregation, stickiness, gelation, dough formation, texturization, mouth-feel
Other	Color, flavor, odor, turbidity

With the trend toward increasing consumption of formulated
convenience foods greater emphasis will be placed on reliable
functional properties in protein ingredients. The idea of producing discrete proteins each with specific functional properties
is being adopted. Currently the relative costs of proteins from
different sources determine their utilization and the more inex-
pensive, but equally functional proteins will be used, nutritional value notwithstanding. Vegetable (oilseed, cereals, leaf) and
yeast proteins should gradually become significant sources of
functional proteins for the food industry in future years as technology develops. The development of appropriate technology under-
lines the need to conduct research to explain the basis of functional properties in proteins. Such information is important in
facilitating the use of new proteins in food applications and de-
signing valid methods for measuring functional properties.

Many processes used for extracting and preparing novel proteins cause denaturation, insolubilization and loss of functional
properties. Because of this the development of practical procedures for the modification of these non-functional proteins to
impart some functional properties is needed. Such modification
can amplify the uses of these protein, facilitate the simulation
of traditional foods, improve compatibility between protein ingredients and aid food manufacturing and processing.

Modification refers to the intentional alteration of the
physical properties of proteins to improve functional properties
(2). The modifying procedure employed may be dictated by the required functionality and ideally it should not result in

destruction of essential amino acids nor formation of potentially
toxic compounds. Methods employed for modification of proteins
include physical, hydrolytic, enzymatic treatments, and chemical
derivatization (Table II).

Table II. Common Methods Used for Protein Modification

Modification	Example
Textural	Extrusion Fiber spinning
Hydrolytic	Alkaline Acid Enzymatic (plastein reaction)
Derivatization	Cations Succinylation Acetylation
Complex Formation	Surfactants Carbohydrate

The development of physical methods for texturizing proteins
has significantly amplified their use in several conventional and
simulated foods. The principal techniques of physical modification which have been thoroughly reviewed are thermoplastic extrusion, fiber spinning and steam texturization (3,4,5).

Partial hydrolysis of proteins using acid, alkali or enzymes
is commonly employed to improve functionality and usefulness of
novel proteins. Acid hydrolysis is the most common method for
preparing hydrolysates of soy, zein, casein, yeast and gluten.
Hydrolysates are used in formulated foods, soups, sauces, gravies,
canned meats, and beverages as flavorants and thickeners (2,3,6).
Alkaline treatments have been employed to solubilize and facilitate protein extraction from soy, single cells, and leaves.
Limited alkaline hydrolysis is used to enhance solubility, emulsifying and foaming properties of proteins (2).

Partial proteolysis has been used by several researchers to
improve functional properties, i.e. foaming, solubility of
proteins (7,8,9). The significant problems associated with enzyme hydrolysis of proteins are excessive hydrolysis occurring
under batch conditions, the generation of bitter flavors during
hydrolysis and the cost of enzymes. Extensive information on
factors affecting proteolysis of proteins and the problem of
bitterness has been reviewed by Fujimaki et al. (7) in conjunction
with studies of the plastein reaction.

Chemical Modification

Chemical derivatization to modify the functional properties
of proteins for food applications has received limited attention
(10,11,12). Intentional chemical alteration of proteins by

derivatization of side chain groups offers numerous opportunities
for the development of new functional protein ingredients.

The conformation of a protein in a particular environment
affects its functional properties. Conformation is governed by
the amino acid composition and their sequence as influenced by
the immediate environment. The secondary, tertiary and quaternary
structures of proteins are mostly due to non-covalent interactions
between the side chains of contiguous amino acid residues. Covalent disulfide bonds may be important in the maintenance of tertiary and quaternary structure. The non-covalent forces are
hydrogen bonding, electrostatic interactions, Van der Waals interactions and hydrophobic associations. The possible importance of
these in relation to protein structure and function was discussed
by Ryan (13).

The current state of knowledge does not permit a correlation
between structure and function though correlations between structural stability and surfactant properties of proteins has been
developed (14). Because of side chains of component amino acids
and their involvement in non-covalent interactions markedly influence structure and function of proteins, the chemical modification of specific reactive groups should facilitate alteration of
functional properties. This approach, though empirical at present, may also provide useful information concerning the role of
specific amino acid groups in conformation and functionality of
protein.

The conformation of most proteins is such that in aqueous
solutions the more polar hydrophilic amino acids, i.e. lysine,
serine, threonine, aspartic and glutamic acids are exposed to the
aqueous environment and the apolar residues are buried in the
interior of the molecule by virtue of hydrophobic interactions.
The modification of the exposed reactive groups, depending upon
the nature of the modifying group may disrupt the non-covalent
interactions, cause alteration in conformation of the protein
with concomitant changes in physical properties. Chemical derivatization may be used to alter the reactivities of food proteins,
to alter their thermal stability or to render them more water
soluble.

Modification of proteins is widely practiced and Knowles (15)
summarized the chemistry of amino acid residue modification that
has been extensively reviewed (16-23). Chemical modification of
proteins, e.g. ε-NH$_2$ group of lysine, predominantly involves
nucleophilic or reductive reactions of the electron rich side
chains (Fig. 1). Reagents susceptible to nucleophilic attack may
react with any of these groups indicating a general lack of specificity. However the reactivity of the group depends upon their
accessibility, the size of the modifying agent and reaction conditions, i.e. pH, temperature and solvent used. Most side chain
groups are modified when they are non-protonated, i.e. nucleophilic. Thus, α-amino and thiol groups are reactive above pH 7.5,
the ε-amino group of lysine above pH 8.5 where they are mostly

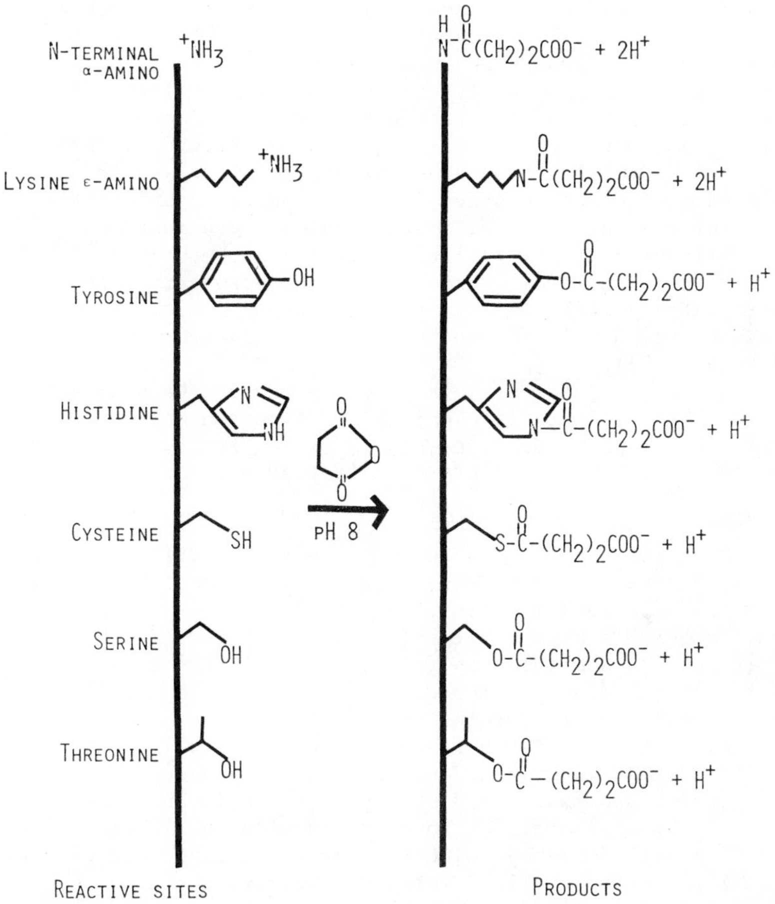

Figure 1. Nucleophilic groups of proteins that are susceptible to acylation. The α-and ε-amino groups are most reactive; the tyrosine phenolic groups generally have a higher pK and are usually more protected from reaction than the amino groups; the histidine and cysteine residues are seldom acylated because the reaction products hydrolyz in aqueous solution and the serine and threonine hydroxyl groups, being weak nucleophiles, are not easily acylated in aqueous solution.

non-protonated and the phenolic group of tyrosine above pH 10
(10,23).

Common methods of derivatization are acylation of amino,
hydroxyl and phenolic residues; alkylation of amino, indole, phenolic, sulfhydryl and thioether groups; esterification of carboxyl;
oxidation of indole, thiol and thioether groups and reduction of
disulfide groups (10). Acylation rates are related to the pK of
the nucleophilic groups being acylated. High reactivity of a
particular group is usually related to low pK values. Acylation
of histidine and cysteine residues is seldom observed because the
reaction products hydrolyze in aqueous solution. Serine and
threonine hydroxyl groups are weak nucleophiles and are not easily
acylated in aqueous solution. Under conditions prevailing in most
foods the ε-amino group of lysine residues are most reactive (24).

Gallenbeck et al. (25) has summarized the work on reductive
alkylation using formaldehyde and sodium borohydride to yield
dimethylated proteins. Oxidation and reduction reactions involving thiol and disulfide groups have been discussed by Ryan (13)
and Feeney (11).

In the remainder of this paper we review the available information on the effects of chemical modification on the functional
properties of plant proteins and report on the use of this approach
for preparing functional proteins from yeast.

Modification of Food Proteins

In relation to food proteins, chemical modification has been
studied for several purposes, i.e. to block reactive groups involved in deteriorative reactions; to improve nutritional properties, to enhance digestibility; to impart thermal stability; to
modify physicochemical properties; to facilitate study of struc-
ture-function relationships and to facilitate separation, processing and refining of proteins (1,2,10,19,20,24).

Numerous undesirable reactions that result in organoleptic,
nutritional and functional deterioration may occur in food proteins during processing and storage. These include the non-
enzymatic or Maillard reactions, transamidation; condensation
reactions with dehydroalanine forming crosslinks, and carbonyl
amine interactions, all of which may involve the free ε-amino
group of lysine (11,23). To minimize these reactions a significant volume of work has been done on the protective modification
of the ε-NH$_2$ of lysine by formylation, acetylation, propionylation
(26) or reductive dimethylation (10,11).

Chemical derivatization of proteins to modify functional
properties has received limited consideration. Cationic derivatives of food proteins are routinely used (e.g. sodium soy isolates and sodium and calcium caseinates) to improve wettability,
dispersibility and handling properties of these proteins (27).
The acylation (and alkylation) of various chemical groups into
proteins provides a method whereby a broad spectrum of functional
compounds may be incorporated into proteins. Several workers have

studied the properties of modified egg proteins (28,29,30); caseins (31,32,33) and fish proteins (34,35). Generally chemical derivatization enhances the thermal stability of food proteins and reduces thermal coagulation and precipitation.

Acetylation of Plant Proteins. Acetic and succinic anhydrides have been the most common derivatives used for the modification of plant proteins and under the conditions employed the ε-NH$_2$ of lysine is the principal site of acylation (Fig. 1). These derivatives are relatively stable requiring strongly acidic conditions for hydrolysis compared to the maleyl derivatives which are acid labile (10).

Acetylation replaces the cationic ε-NH$_2$ of lysine by neutral acetyl groups. This reduces electrostatic attractions between adjacent polypeptides with a resultant 'loosening' of interfolded peptides; it causes a shifting of isoelectric point to lower pH values and slightly enhances solubility in the acidic pH range. Reduction in electrostatic attraction between proteins may reduce the gelling ability of proteins upon heating. Acetylation protects the ε-NH$_2$ group of lysine from participating in unwanted deteriorative reactions and upon digestion the acetyl group can be hydrolyzed by renal deacylase (26).

Acetylation of soy protein causes dissociation of the 11S component with a concomitant increase in 7S and 2S components (36) indicating that the cationic ε-NH$_2$ group of lysine was important in stabilizing the 11S polymer of glycinin via electrostatic forces, though the introduction of the bulkier acetyl group may also disrupt adjacent hydrogen bonds. Acetylation decreased water binding by 20% (0.4 vs. 0.5 g/g) and caused a slight reduction in water activity of soy protein. The loss of water sorption was attributed to elimination of the charged ε-amino group of lysine (36). Acetylation increased bulk density, improved wettability, increased solubility, and reduced the isoelectric point to pH 4 (36,37). The viscosity of soy protein dispersions (36) was increased presumably as a result of disaggregation of the protein. Acetylation eliminated the capacity of soy proteins (20% dispersions) to form gels upon heating, indicating that electrostatic bonds were important in gelation. Acetylation reduced emulsifying properties compared to the unmodified soy isolate but had negligible effects on foaming capacity (37).

Acetylated cottonseed protein demonstrated significantly higher water and oil holding capacities and improved foaming properties (38) compared to unmodified proteins (Table III). Thus, while acetylation does not significantly enhance functional properties of proteins, it improves thermal stability and since acetylated proteins are susceptible to enzyme hydrolysis in vivo it affords a useful reagent for protection of ε-NH$_2$ groups of lysine (11).

Table III.   Functionality of Acylated Cottonseed Products

	Cottonseed Flour		
Capacity	Glandless	Acetylated	Succinylated
	- - - - - - - - (ml/g) - - - - - - - - - -		
Water-holding	3.50	7.87	5.53
Oil-holding	2.60	8.97	19.60
Emulsifying	456.67	416.67	776.67
Foam	126.67	250.00	150.00

Succinylation

Succinic anhydride reacts with free amino groups, the phenolic and thiol groups of tyrosine and cysteine. The latter esters
spontaneously hydrolyze in aqueous solution (39). The ε-amino
group of lysine reacts most readily. Succinylation has three
major effects on the physical character of proteins; it increases
net negative charge, changes conformation and increases the pro-
pensity of proteins to dissociate into subunits (39,40,41,42).

The introduction of succinate anions alters the conformation
of protein and penetration of water molecules is easier because of
the loosened state of polypeptides. The structural instability of
succinylated proteins is caused by the high net negative charge
and the replacement of short range attractive forces in the native
molecule with short range repulsive ones with subsequent unfolding
of polypeptide chains. This probably accounts for the looser texture, higher bulk density, lighter color, and enhanced solubility
of succinylated food proteins. The effects of succinylation on
plant proteins are summarized (Table IV).

Succinylation of a protein, unless all component ε-NH$_2$ groups
are fully acylated, results in a more heterogeneous population of
proteins (43,44,45). This is due to variation in the extent of
succinylation of protein components, but is also caused by extensive dissociation of the derivatized proteins. Succinylation to
different degrees caused the progressive dissociation of arachin
into subunits (45). The unmodified protein gave a single symmet-
rical peak with a sedimentation coefficient of 14S but as the protein was progressively succinylated the 14S peak decreased with a
concurrent increase in the 9S and 4S components, i.e. at 60%
succinylation, the 9S peak accounted for about 40% of the total
protein. The 9S component decreased as succinylation was increased
to a maximum of 80% of all ε-NH$_2$ groups, at which state only the
4S component was observed. Thus succinylation caused dissociation
of arachin [14S ⇌ 9S ⇌ 4S] into low molecular weight proteins.
Concurrently the viscosity of arachin dispersions increased with
succinylation, about 3.5 fold at 80% succinylation (Fig. 2). From
the sedimentation patterns a decrease in viscosity would be anticipated due to dissociation of the components, however, the degree
of unfolding of the dissociated components and the associated increase in hydrodynamic volume more than counterbalanced the

Table IV.   The Effects of Acylation on Some Functional Properties
of Plant Proteins

Protein	Modifica-tion	Change in Properties	Refer-ence
Peanut	Succinyl-ation	Functional - increased nitrogen solubility in acidic pH, water-absorption capacity & viscosity	(43)
		Physico-chemical - dissociation into subunits, increased viscosity due to swelling, increased random-coiled structure, change in spectral properties	(45)
Cotton-seed	Succinyl-ation	Caused increase in specific viscosity, water-holding, oil-holding and foaming capacities	(38)
Soybean	Alkali (pH > 10)	Increased dispersibility, solubility, resistance to aggregation (heat, etc.), elasticity (better fiber formation)	(91)
	Succinyl-ation	Increased solubility at acidic pH, increased tolerance to Ca^{2+}, good foam capacity, stability, emulsifying activity and emulsion stability. More resistance to aggregation. Lowered viscosity	(12)
	Reduction (sulfite)	Reduced viscosity of water dispersion & increased viscosity in salt solution and resistance to aggregation	
Wheat	Acylation	Increased nitrogen extractability and viscosity. Dissociation into subunits	(44)
Leaf	Succinyl-ation	Increased bulk density, solubility, foaming capacity. Improved flavor. Enhanced emulsifying activity	(12)
Yeast	Acylation	Lower digestibility with pepsin and pancreatin, decreased emulsion stability, increased viscosity, altered solubility	(61, 88)

expected apparent decrease in viscosity due to dissociation. Progressive succinylation caused conformational changes and unfolding of the polypeptides as indicated by the changes in absorption spectra (blue shift of tyrosine peak) and increased specific ellipticity of the dichroic spectra (45) around 212 nm. Grant (46) reported that succinylated wheat flour proteins produced a heterogeneous population of derivatized proteins that were 90% soluble in water. The succinylated derivatives extensively dissociated in solution compared to untreated proteins.

Succinylation substantially increases specific volume of soy and leaf proteins (12,37). The succinylated soy protein becomes very fluffy and the color becomes much lighter, changing from a tan to a chalk white as the extent of derivatization is increased (12,47). No odors nor flavors were imparted by the succinylation process. Succinylation improved the whiteness and dispersibility characteristics of soy protein making it suitable for incorporation into coffee whiteners (47). Succinylated soy proteins hydrate rapidly on the tongue, taste clean, but slightly acidic. It is not known if derivatization facilitates the removal of off-flavors from modified proteins.

Succinylation significantly enhances the rate of hydration of soy, peanut and cottonseed proteins (12,38,43). It causes a marked improvement in the water solubility of soy protein and also of leaf protein concentrate (12). It decreased the isoelectric point of both soy and peanut proteins by approximately 0.5 pH unit (from pH 4.5 to 5.0); significantly enhanced solubility between the isoelectric point and pH 6, but progressively reduced solubility of both soy and peanut protein below the isoelectric point (12,43).

Succinylation of proteins converts the cationic ε-amino groups to anionic residues and the increased negative charge alters the physicochemical character of the protein resulting in enhanced aqueous solubility and changes in other physicochemical properties. Thus, in the unmodified soy isolate electrostatic attractions between neighboring ammonium and carboxyl groups enhance protein-protein interactions which lowered solubility (41,45). Following succinylation the ammonium cations of lysine are replaced by succinate anions. Electrostatic repulsions occur between the added carboxyl groups and the neighboring native carboxyl groups producing fewer protein-protein interactions and more protein-water interactions which enhance hydration and aqueous solubility. Since net negative charge is proportional to the extent of amino succinylation, the enhancement in aqueous solubility was positively related to the extent of derivatization.

Succinylation increased the emulsifying activity and emulsion stability of soy and cottonseed proteins (12,38). The pH emulsifying capacity profiles of succinylated protein paralleled those of the solubility curves and in all cases the succinylated protein had about double the emulsifying capacity of the unmodified

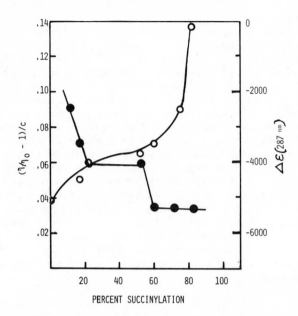

Figure 2.   The effects of progressive succinylation on the viscosity $(\eta/\eta_0 - 1)/C)$ and uv absorption, i.e. tyrosine ($\Delta\epsilon$ 287) of peanut protein dispersions

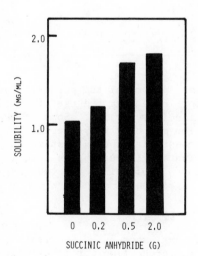

Figure 3.   The effect of succinylation on the solubility of soy protein isolate

protein (Fig. 3).

The foaming capacity of succinylated soy protein was significantly better than those of the unmodifided proteins. Foam
volumes progressively increased with pH from 3 to 10 (12). Succinylation caused a small increase in foaming capacity of cottonseed
flour (38). Solubility is required for the production of protein
foams (48), and succinylation substantially increased the foaming
ability of soy isolate by enhancing their solubility.

Though succinylation causes dissociation of proteins into
smaller molecular sizes it usually is associated with an increase
in viscosity of dispersions of modified protein (42,44,45,49).
The viscosity of modified peanut protein increased with the extent
of succinylation (Table V) and this was most pronounced at high
concentrations of the protein (43,45). Despite the increased
electronegativity of succinylated proteins the addition of calcium
to dilute dispersions resulted in no apparent increase in viscosity (12,37). Melnychyn and Stapely (47) noted a reduced viscosity
in succinylated vegetable proteins. They noted the thermal stability of these proteins when heated to 100°C and suggested their
use for coffee whiteners.

Table V.   Effect of Degree of Succinylation and Protein Concen-
           tration on Viscosities of Peanut Flour Dispersions

Degree of	Protein concentration (mg/100 ml)			
Succinylation	1	2	5	10
	- - - -Viscosity (CP) of dispersions- - - - -			
0	3.5	13.0	14.1	22.4
10	3.7	16.0	18.2	35.7
40	3.5	16.6	19.4	46.1
70	4.0	16.8	20.6	45.8
100	4.4	16.4	23.1	47.0

from Beuchat (77)

The limited studies have revealed some of the potential benefits of chemical derivatization for increasing the use of novel
proteins by expanding their functional properties. Some of the
properties imparted by succinylation may have very specific or
unique applications, e.g. as thermo-stable protein dispersions in
coffee whiteners, beverages. Derivatization can be used to impart
functionality into denatured, insoluble proteins, and thereby increase their value as functional ingredients.

YEAST PROTEIN ISOLATION

Because of the need for additional sources of nutrients and
the constraints on conventional crop production, it is expedient
to develop supplementary sources of food protein, e.g. yeast
protein. The most popular yeasts belong to the general Saccharomyces and Candida because of their long tradition of use in foods.

The general digestibility of yeast is good but several problems, i.e. potential toxicity (acute and chronic), gastrointestinal problems, nausea, vomiting, diarrhea, rashes, caused by yeast constituents (endotoxins, metabolites) emphasize the general need for refining yeast protein for human diets.

Several of the problems associated with the consumption of microbial cells are inherent to the cell and not intrinsic properties of the proteins per se. The presence of cell wall materials in unfractionated cells is undesirable because it reduces the bioavailability of the proteins; may contain antigenic, allergenic agents and factors causing nausea, gastrointestinal disturbances (flatulence diarrhea) and the cell wall materials also cause darkening of the cell material (50,51,52,53,54).

Another significant problem associated with consumption of microbial cells is their high content of nucleic acid (NA) which ranges from 8 to 25 gms nucleic acid per 100 gms protein and most of the nucleic acid is present as RNA (55). Before single-cell protein can be used as a major source of protein for human consumption, the content of nucleic acid has to be reduced, so that the daily intake of nucleic acid from yeast would not exceed 2 gms (i.e. 20g yeast). Higher quantities cause uricemia and continued ingestion of SCP may result in gout (56,57).

Ideally microbial cells should be consumable directly as food or food ingredients. However, because of their nucleic acid content; the presence of undesirable physiologically active components; the deleterious effects of cell wall material on protein bioavailability and the lack of requisite and discrete functional properties, rupture of cells and extraction of the protein is a necessary step. Importantly, for many food uses (particularly as a functional protein ingredient) an undenatured protein is required. For these reasons and for many potential applications of yeast protein(s) it is very desirable to separate cell wall material and RNA from the protein(s) for food applications. Much research is needed to develop a practical method for isolation of intact, undenatured yeast proteins from the yeast cell wall material to ensure the requisite nutritional and functional properties.

The successful adoption of SCP will depend upon their acceptance by the food manufacturing industry. To achieve adoption (in addition to availability, cost, nutritive value, safety) the physicochemical properties and processing characteristics of the protein(s) must be fully described. The latter, functional properties, must be known to determine (and predict) how these proteins will behave in a variety of food applications, if they can be used to replace more expensive proteins, and their capacity for the fabrication of new proteinaceous foods.

Intact dried microbial cells have limited functional properties. Labuza et al. (58) reported that spray drying of yeast cells at higher temperatures improved the functional performance of yeast cells in bread making but they were still inferior to control samples. Spray drying of yeast cells at 75 and 124°C

caused insolubilization of 72 and 84% of the protein, respectively
(59).

Thus, to improve nutritional value and exploit functional
properties, extraction and concentration of the protein is necessary. Numerous techniques have been developed for the extraction
of yeast proteins (50,51,60,61,62,63) but few produce a protein
with the properties that meet the necessary nutritional and functional criteria. For extraction of protein from microbial cells,
chemical treatment or physical rupture is necessary to render the
cell contents available to the extractant (61-68). For commercial
processing, efficient and complete cell breakage is highly desirable for ensuring maximum extractability and recovery of protein.
Furthermore, for many food applications it is required that denaturation (thermal, surface) of the protein is minimized to ensure retention of important functional properties. Mechanical
methods require large energy expenditures to achieve efficient
rupture of the cell wall and thermal denaturation of the protein
during cell breakage is frequent.

A common problem associated with rupture of yeast cells and
protein extraction is proteolysis. Yeast cells contain a full
complement of intracellular proteolytic enzymes which may be liberated after the cells are broken either by autolysis or by mechanical disruption. These liberated proteolytic enzymes, unless inactivated during the isolation and purification of yeast proteins,
hydrolyze the proteins causing poor yields of intact protein (55,
69,70).

The exploitation of ribonuclease (66,71,72) in depleting
nucleic acids involves the incubation of the extracted protein
(containing the cellular RNA) around neutral pH at 55°C for one or
two hours to activate the RNase and hydrolyze the nucleic acids.
Unfortunately, these conditions also result in the activation of
the proteolytic enzymes of yeast cells. These concurrently hydrolyze the proteins to low molecular weight peptides and significantly reduce the yield of intact proteins.

For maximizing protein yield and minimizing contaminant nucleic acids, alkali extraction at elevated temperatures is a feasible procedure (63,73). However denaturation of proteins during
extraction is a serious problem because it significantly destroys
functional properties and limits the food uses of the extracted
protein (2, 74). In addition to denaturation, exposure of protein
to alkaline treatments may also cause other undesirable effects,
i.e. racemization of amino acids, β-elimination and crosslinking
or certain amino acids and formation of potentially antinutritive
compounds (11,23,75).

In conjunction with research on protein extraction from yeast,
we investigated methods for the maximum recovery of protein possessing good functional properties but low in nucleic acid.
Therefore, we examined the feasibility of making the yeast protein
resistant to proteolysis during extraction and nucleic acid reduction. Using established extraction procedures (76), we observed

that the addition of succinic anhydride to dispersions of mechanically disrupted yeast cells at pH 8.5 resulted in a significant
increase in protein extractability. Whereas only 55-60% of protein was extracted from mechanically disrupted cells at pH 8.5,
succinylation of yeast cell homogenate resulted in extraction of
90% of the yeast protein (Fig. 4). This was significant because
the pH of extraction represented very mild conditions compared to
other conventional procedures.

Incubation of succinylated yeast homogenate at 55°C (pH 6.0).
to activate endogenous ribonuclease and hydrolyze the contaminant
RNA resulted in the significant reduction of the extensive proteolysis normally observed (Table VI). Furthermore, the coagulation and precipitation of protein that normally occurs during this
step was also eliminated.

Table VI. The Effect of Succinylation on Proteolysis of Yeast
           Proteins by Endogenous Proteases During Incubation
           (pH 6.0, 55°C) for RNA Reduction

| Incubation | Yield of Protein | |
Period	Unsuccinylated	Succinylated
0	67.4	83.1
1/2 hr.	59.0	77.7
1 hr.	53.7	79.5
2 hr.	42.5	71.1
3 hr.	44.9	72.3
5 hr.	30.0	72.3

Thus, succinylation prevented the aggregation and coagulation of
yeast proteins normally observed during RNA hydrolysis with endogenous ribonuclease and markedly increased the recovery of intact,
soluble protein. Lindbloom (69) reported that up to 70% of yeast
protein may be degraded during incubation to reduce RNA (pH 6,
55°C, 5 hr). However, in our studies, when the protein was succinylated proteolysis was inhibited compared to the non-succiny-
lated control. This inhibition may have resulted from inactivation of the proteases by succinylation; non-susceptibility of
succinylated protein to protease attack, or the inhibition of
proteases by succinylated hydrolytic products.

Because derivatization impaired proteolysis, its effect on
endogenous ribonclease was studied. Ribonuclease activity was inhibited with increasing succinylation and dropped sharply at an
anhydride to protein ratio of 0.8 where 80% of ε-NH$_2$ groups of the
total protein were succinylated (Fig. 5). Because the ribonuclease
was inactivated, an increased nucleic acid content in the precipitated proteins was expected. However, we observed less nucleic
acid (NA) in precipitated succinylated protein than in the non-
modified controls. Maximum precipitation of succinylated protein
occurred around pH 4.5 and contained only 1.8% nucleic acid on a

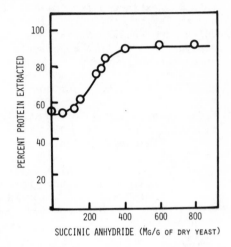

Figure 4. The increased quantity of protein extracted from homogenized yeast cells at pH 8.5 following succinylation

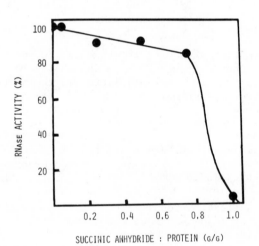

Figure 5.   The effects of succinylation during extraction of yeast protein on the endogenous ribonuclease activity (pH 6.0, 55°C)

dry weight basis whereas protein precipitated from unsuccinylated
yeast extract contained 28% NA (Table VII). Succinylation of the
protein by increasing its net negativity accentuated the small
differences in the total charge between the derivatized proteins
and the NA thereby facilitating the almost exclusive precipitation
of protein at pH 4.5 (Fig. 6). This procedure greatly simplifies
the separation of NA from yeast protein which can be recovered in
high yields in a highly soluble state.

Table VII.   The Influence of pH of Protein Precipitation on the
             Content (%) of Nucleic Acid in Nonsuccinylated and
             Succinylated Yeast Protein

pH of Precipitation	Nucleic Acid Content of Protein (%)	
	Nonsuccinylated	Succinylated
2	38	19
3	33	17
4	30	2
4.5	28	1.8
5.0	28	2.5

   The procedure described above may be a significant and practical step in developing a process for the isolation of a soluble
protein, low in nucleic acid and in high yields from yeast cells
(Fig. 7). The procedure is rapid, it eliminates the incubation
(4-5 hr) step, avoids proteolysis, and is amenable to current
extraction procedures. The method is effective when dicarboxylic
residues are introduced but is less successful with acetylation.
Furthermore, derivatization with certain cyclic dicarboxylic anhydrides is a reversible process. This procedure provides an example of chemical derivatization facilitating the isolation of
protein.

Functional Properties of Yeast Proteins

   Yeast proteins, in addition to being nutritionally good and
safe, must possess functional properties to be adopted by the food
industry and to gain general consumer acceptance. Information on
functional properties of SCP is needed to evaluate and predict how
these new proteins behave in specific food systems and if they can
be used to complement, replace or simulate conventional proteins
in different foods. There has been a surprisingly limited amount
of information published concerning the functional properties of
yeast proteins intended for food applications, though proprietary
knowledge exists (74,77,78,79,80,81,90).
   Flavor, color and texture are important primary properties of
proteins. In novel proteins the absence of flavors or odors is
desired to render the new protein compatible with the food to
which they are added. Off-flavors frequently limit the use of
yeast protein preparations (81). Frequently these arise from the

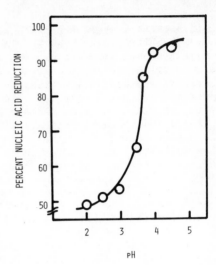

Figure 6. *Influence of pH of precipitation on the extent of nucleic acid reduction in precipitated succinylated yeast proteins*

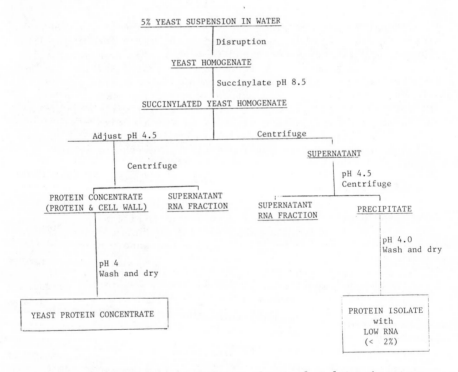

Figure 7. *Outline of the succinylation procedure for the isolation of yeast protein concentrate (72% protein) or yeast protein isolate (92% protein) with low (< 2%) nucleic acid*

lipid materials which are associated with the isolated proteins.
     Solubility is a critical functional characteristic because
many functional properties depend on the capacity of proteins to
go into solution initially, e.g. gelation, emulsification, foam
formation. Data on solubility of a protein under a variety of
environmental conditions (pH, ionic strength, temperature) are
useful diagnostically in providing information on prior treatment
of a protein (i.e. if denaturation has occurred) and as indices of
the potential applications of the protein, e.g. a protein with
poor solubility is of little use in foams). Determination of
solubility is the first test in evaluation of the potential functional properties of proteins and retention of solubility is a
useful criterion when selecting methods for isolating and refining
protein preparations (1).   Several researchers have reported on
the solubility of extracted microbial proteins (69,82,83,84). In
many instances yeast proteins demonstrate very inferior solubility
properties below pH 7.5 because of denaturation.
     The foaming and emulsifying capacity of yeast proteins have
been studied (83,84).   Rha et al. (85) have summarized the limited
information on the spinning, i.e. fiber forming ability of proteins
from C. utilis.
     The limited data indicate that yeast proteins extracted by
the current, more conventional methods lack the requisite functional properties for many applications.

Modification

     During isolation by conventional methods yeast proteins fre-
quently become denatured, insolubilized and display poor functional
properties.   These proteins can be rendered more soluble by
limited hydrolysis with acid, alkali or proteolytic enzymes.   Protein hydrolyzates are most commonly prepared by partial acid
hydrolysis and yeast hydrolyzates are popular as food flavorings
and ingredients (66).   Acid hydrolyzates have flavors resembling
cooked meats and are widely used by canners to impart brothy,
meaty flavors to soups, gravies, sauces, canned meats.
     Alkali treatment has been used to improve the functional
properties of the insoluble protein prepared by heat precipitation
of an alkaline extract of broken yeast cells (63). Heating yeast
protein at pH 11.8 followed by acid precipitation (pH 4.5) yielded
a preparation composed of polypeptides with increased aqueous solubility.   It also increased foaming capacity of the protein 20-
fold.   The emulsifying capacity of the modified protein was good
whereas the original insoluble protein was incapable of forming an
emulsion.   Alkali treatment must be carefully controlled to avoid
its possible deleterious effects (24,75), e.g. alkaline treatment
of yeast protein resulted in a loss (60%) of cysteine (63).
     Enzyme hydrolysis is occasionally used to modify the func-
tional properties of proteins and yeast autolyzates are used
commercially as food flavorants (66,86).   Partial proteolysis of

novel proteins improves their solubility and foaming properties
(87). The problems associated with enzyme hydrolysis are the
generation of bitter peptides and the cost of enzymes. The plastein reaction effectively debittered pepsin hydrolyzates of yeasts
(Saccharomyces, Candida); aided the release of lipid materials responsible for the development of off-flavors and reduced pigments
and heme materials present in the original SCP (7).

Chemical modification of yeast protein has received limited
attention though as described above it has potential as a method
for facilitating recovery of yeast protein. Current studies are
concerned with determination of the functional properties of proteins succinylated during the extraction. The composition of
yeast proteins prepared by different methods is shown (Table 8).
Noteworthy is the protein and nucleic acid concentration in the
yeast isolate which differed from the concentrate in that cell
wall material was removed by centrifugation.

Table VIII.   Composition of Yeast Cells, Succinylated Yeast Protein
              Concentrate and Succinylated Yeast Protein Isolate
              (expressed as g/100g)

Sample	Nucleic Acid	Protein Nitrogen	Carbo-hydrate	Crude Lipid
Yeast Cell Homogenate	7.90	6.71	35.40	2.38
Succinylated Yeast Homogenate (Pro-tein Concentrate)	2.0	11.28	24.80	4.45
Succinylated Yeast Protein Isolate	1.80	15.32	7.0	3.47

The absorption spectra of three yeast protein preparations
prepared by different procedures were compared (Fig. 8). The
presence of nucleic acid which has a λ maximum at 260 nm tend to
shift the absorption spectrum of yeast protein to lower wave-
lengths. The ratio of absorption at 280 to 260 nm is indicative
of NA contamination in protein samples; a ratio of more than one
indicates pure protein devoid of nucleic acid whereas a ratio of
0.65 indicates approximately 30% contamination with NA. The
yeast protein extracted with alkali and directly acid precipitated
showed a λ max at 260, a 280/260 ratio of 0.67 and contained 28%,
NA determined chemically. Protein extracted in alkali, adjusted
to pH 6 and incubated at 55°C for 3 - 5 hours, to reduce NA with
endogenous ribonuclease, had a λ max at 260, a 280/260 ratio of
0.8 and a NA content of 3.3% while yeast protein prepared by the
succinylation procedure and precipitated at pH 4.5 showed a λ max
at 275 nm, a 280/260 ratio of 1.0 and nucleic acid content of 1.8.

Succinylation caused dissociation of the yeast proteins.
This was difficult to detect with methods that separate proteins

on the basis of molecular size. Using gel filtration on Sephadex
G-100 yeast protein gave two peaks, the major peak (∿80%) emerging
in the void volume corresponded to proteins exceeding 100,000
daltons (76) and a second peak, probably contained a heterogeneous
group of proteins with a lower range of molecular weights. The
succinylated protein also gave two peaks with the major peak emerging immediately after the void volume and the second minor peak.
This pattern was not indicative of dissociation of the succinylated protein. However, sedimentation velocity studies of the
succinylated yeast proteins gave a single peak with a sedimentation coefficient of approx. 1S indicating the extensive dissociation of the larger protein components of yeast. Therefore, the
pattern obtained upon Sephadex filtration, i.e. the major peak
emerging after the elution volume, is probably explained by the
significant increase in the hydrodynamic volume of the dissociated
succinylated yeast proteins caused by unfolding of the molecules.
Attempts to resolve dissociated succinylated yeast proteins by gel
electrophoresis were unsuccessful because the proteins failed to
enter the gel. This dissociation caused by succinylation usually
causes a marked increase in viscosity of these preparations.

Succinylation reduced the isoelectric point of yeast proteins
from pH 4.5 to 4.0 and markedly improved their solubility in pH
range 4.5 to 6. Heat denatured yeast proteins were facilely solu-
bilized following succinylation (74). Succinylated yeast proteins
were very stable to heat above pH 5, and remained soluble at temperatures above 80°C. As the degree of succinylation was increased
the rate of precipitation of the derivatized protein increased in
the neighborhood of the isoelectric point and much larger protein
flocs were obtained facilitiating their recovery (74).

The emulsifying capacity of the yeast proteins was progressively improved with the extent of succinylation (Table IX) as
measured by the turbidimetric technique (89). The modified yeast
proteins had excellent emulsifying activities compared to several
other common proteins. McElwain et al. (88) observed that succinylation of yeast protein increased emulsion viscosity but decreased emulsion stability.

Table IX. Emulsifying Activity Index Values of Succinylated Yeast
           Protein

Sample	Succinylation (%)	EAI (m^2, g^{-1})
1	0	20
2	24	70
3	13	90
4	28	220
5	44	221
6	54	328
7	88	356

EAI denotes surface area of emulsion formed per gram protein (89).

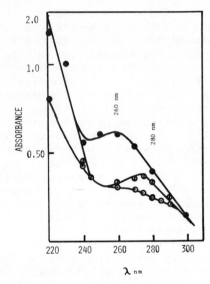

Figure 8. Absorption spectra of yeast proteins prepared by three different methods.

Legend: crude protein prepared by precipitation of an alkali extract at pH 4.0 (●); yeast protein obtained following activation of endogenous ribonuclease (82) (☉), and yeast protein prepared by the succinylation procedure (○).

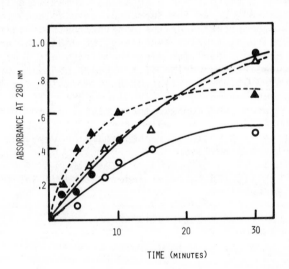

Figure 9. Rate of hydrolysis of succinylated and nonsuccinylated yeast proteins by α-chymotrypsin, (○) unsuccinylated and (●) succinylated, and trypsin, (△) unsuccinylated and (▲) succinylated

Succinylation during extraction significantly improved the foaming capacity of yeast protein. The succinylated protein was equivalent to ovalbumin whereas unmodified denatured yeast protein was very inferior and failed to retain gas efficiently during expansion. Maximum foam strength of succinylated yeast protein and ovalbumin occurred slightly above the isoelectric points of these two proteins. The ovalbumin foam was significantly stronger than the succinylated yeast protein at similar concentrations of protein. The foam stability of succinylated yeast protein was only slightly inferior to that formed by egg albumin.

The in vitro digestibility of succinylated yeast proteins by α-chymotrypsin and trypsin was compared (Fig. 9). In both instances hydrolysis of both derivatized and unmodified yeast protein was quite similar, though the α-chymotrypsin hydrolyzed the succinylated protein more rapidly.

Conclusions

Although chemically modified proteins may not be permitted in foods for some time, basic research in this area should continue and not be restricted by current economic and/or regulatory criteria, even though these two factors may ultimately determine the application of modified proteins in foods. The limited information already available has revealed the potential of derivatized proteins in expanding the use of novel proteins with limited functional properties. More extensive knowledge is required before conclusive judgments can be made concerning modification of food proteins. While nutritional value and safety are of paramount importance, appropriate functional properties are the key attributes in the assembly, fabrication and processing of the heterogeneous range of foods required by different consumers throughout the world.

Research is needed to develop functional reagents that are safe and do not impair the availability of nutrients; that possess good specificity for particular nucleophilic groups on the protein and that are readily hydrolyzed. The nutritional impact and safety of chemically modified proteins should be adequately tested for each derivative. However, it is unlikely that modified proteins would become significant sources of dietary protein for any particular population group. Research should be continued even though modification may reduce the availability of some amino acid residues because modified functional proteins may well facilitate the incorporation of nutritionally superior proteins into a variety of foods appealing to a broad range of consumers. Research to develop functional derivatives that specifically react with non-limiting amino acids should be encouraged.

Chemical modification may help researchers to elucidate the physicochemical basis of protein functionality, and understanding the physical basis of specific functional properties should enable scientists to design better functional derivatives. Reagents that

cause unfolding of polypeptides in a particular manner could be
used to enhance surfactant properties and derivatization with
polyhydroxy compounds might enhance water sorption properties.
It is very conceivable that series of protein derivatives with
graded differences in functional properties (thermal stability,
hydrophobicity, hydrophilicity, molecular dimensions, etc.) will
become available in the future. Such functional derivatives
should facilitate the exploitation of the wide range of novel proteins in the world.

Finally, in conjunction with research on modification there
must be a concomitant and commensurate research effort to develop
standardized quantitative methods for measuring functional properties of conventional and modified proteins.

Acknowledgement. This work was partly supported by Grant No. Eng.
75-17273 from the National Science Foundation.

Abstract

To achieve success as protein ingredients for food formulation and fabrication, novel proteins should possess a range of
functional properties. Frequently during extraction, refining and
drying, plant and yeast proteins, intended for food uses, become
denatured or altered and subsequently display poor functional
properties which render them of limited use. Chemical modification provides a feasible method for improving the functional properties of plant and yeast proteins and potentially may make it
possible to tailor proteins with very specific functional properties. In this review the information on modified plant proteins
is reviewed and the use of succinylation for the recovery of
yeast proteins with low nucleic acid is described.

Literature Cited

1. Kinsella, J. E. Crit. Rev. Food Sci. & Nutr. (1976) 7, 219.
2. Kinsella, J. E. Chemistry and Industry (1977)5, 177.
3. Kinsella, J. E. in Handbook Food and Nutrition (1978) in
    press.
4. Horan, F. E. in New Protein Foods, Vol. 1A., A. Altschul, Ed.,
    Acad. Press, N. Y. (1974).
5. Gutcho, E. Textured Food and Allied Products, Noyes Data Corp.
    Park Ridge, N.J. (1973)
6. Connell, J. Can. & Food Ind. (1966) 37, 23.
7. Fujimaki, M., Arai, S., Yamashita, M. in Food Proteins.
    R. Feeney and J. Whitaker, Eds. Am. Chem. Soc., Washington,
    D.C. (1977) pg. 156.
8. Whitaker, J. R. in Food Proteins. R. Feeney and J. Whitaker,
    Eds., Am. Chem. Soc., Washington, D.C. (1977) pg. 14.
9. Hermansson, A. M., Olsson, I., Holunberg, B. Lebens-Wiss u
    Technol. (1976) 7, 176.

10. Feeney, R. E. in Food Proteins. R. Feeney and J. Whitaker, Eds. Am. Chem. Soc., Washington, D.C. (1977) pg. 3.

11. Feeney, R. E. in Evaluation of Proteins for Humans. C. Bodwell, Ed. AVI Publ. Co., Westport, Conn. (1977) pg. 233.

12. Franzen, K. M., Kinsella, J. E. J. Agr. Fd. Chem. (1976) 24, 788.

13. Ryan, D. S. in Food Proteins. R. Feeney and J. Whitaker, Eds. Am. Chem. Soc., Washington, D.C. (1977) pg. 67.

14. Phillips, M. C. Chem. and Industry (1977) March, pg. 170.

15. Knowles, J. R. in Chemistry of Macromolecules. H. Gutenfreund, Ed. University Park Press, Baltimore, Md. (1974) pg. 149.

16. Singer, S. J. Advan. Protein Chem. (1967)22, 1.

17. Cohen, L. A. Ann. Rev. Biochem. (1968) 37, 695.

18. Vallee, B. L., Riordan, J. Ann. Rev. Biochem. (1969) 38, 733.

19. Stark, G. R. Advan. Protein Chem. (1970) 24, 261.

20. Means, G., Feeney, R. E. Chemical Modification of Proteins. Holden-Day, San Francisco (1971).

21. Hirs, C. M., Timasheff, S. Methods in Enzymology, Vol. 25. Academic Press, N. Y. (1972).

22. Knowles, J. R. Phil Trans. Roy. Soc. Land (1970) B257, 135.

23. Friedman, M. Protein Cross-Linking: Biochemical and Molecular Aspects. M. Friedman, Ed., Advan. Exptl. Med. & Biol. (1977) 86A. Plenum Press, N. Y.

24. Friedman, M. in Food Proteins. J. Whitaker and S. Tannenbaum, Eds. AVI Publ. Co., Westport, Conn. (1977) pg. 446.

25. Galembeck, F., Ryan, D. S., Whitaker, J., Feenery, R. E. J. Agr. Food Chem. (1977) 25, 238.

26. Bjarson, J., Carpenter, K. J. Br. J. Nutr. (1970) 24, 213.

27. Thomas, M. A., Baumgartner, P., Hyde, K. Aust. J. Dairy Tech. (1974) 29, 59.

28. Oppenheimer, M., Barany, K., Harmour, G., Genton, J. Arch. Biochem. Biophys. (1967) 120, 108.

29. Gandhi, S., Schultz, J., Boughey, F., Forsythe, R. J. Food Sci. (1968) 33, 163.

30. Evans, M., Irons, L. German Patent 1,951,247 (1971).

31. Hoagland, P. D. Biochemistry (1968) 7, 2542.

32. Evans, M., Irons, L., Petty, J. M. Biochim. Biophys. Acta. (1971) 243, 259.

33. Creamer, L., Roeper, J., Lohrey, E. N.Z.J. Dairy Sci. Technol. (1971) 6, 107.

34. Groniger, H. J. Agric. Food Chem. (1973) 21, 978.

35. Chen, L., Richardson, T., Amundson, C. J. Milk Food Technol. (1975) 38, 89.

36. Barman, B. G., Hansen, J. R., Mossey, A. R. J. Agr. Food Chem. (1977) 25, 638.

37. Franzen, K. M., Kinsella, J. E. J. Agr. Food Chem. (1976) 24, 914.

38. Childs, E. A., Park, K. K. J. Food Sci. (1977) 41, 713.

39. Gounaris, A., Perlmann, G. J. Biol. Chem. (1967) 242, 2739.

40. Habeeb, A.F.S.A. Arch. Biochem. Biophys. (1967) 121, 652.

41. Habeeb, A.R.S.A., Cassidy, H., Singer, S. Biochim. Biophys. Acta. (1958) 29, 587.

42. Girdhar, B. K., Mansen, P. M. J. Food Sci. (1974) 39, 1237.

43. Beuchat, L. J. Agr. Food Chem. (1977) 25, 258.

44. Grant, D. Cereal Chem. (1973) 50, 417.

45. Shetty, K. J., Rao, N. S. Int. J. Peptide Protein Res. (1978) 11, 305.

46. Donovan, J. W. in Physical Principles and Techniques of Protein Chemistry. S. J. Leach, Ed., Academic Press, N.Y. (1969) pg. 107.

47. Melnychyn, P., Stapley, R. U.S. Patent 3,764,711 (1973).

48. Eldridge, A., Hall, P., Wolf, W. Food Technol. (1963) 17, 1592.

49. Chen, L. F., Richardson, T., Amundson, C. H. J. Milk Food Tech. (1975) 38, 89.

50. S. Tannenbaum, R. Mateles, Eds. Single Cell Protein. M.I.T. Press, Cambridge, Mass. (1968).

51. Tannenbaum, S., Wang, D. I., Eds. Single Cell Protein II, M.I.T. Press, Cambridge, Mass. (1973).

52. Young, V., Scrimshaw, N. in Proteins in Human Nutrition. C. Bodwell, Ed., AVI Publ. Co., Westport, Conn. (1977) pg. 11.

53. Scrimshaw, N. S. in Single Cell Proteins II. S. Tannenbaum, Ed., M.I.T. Press, Cambridge, Mass. (1973) pg. 46.

54. Young, V., Scrimshaw, N. S., Milner, M. Chem. & Industry (1976) July pg. 588.

55. Sinskey, A. J., Tannenbaum, S. R. in Single-Cell Proteins II. S. R. Tannenbaum and D.I.C. Wang, Eds. M.I.T. Press, Cambridge, Mass. (1975) pg. 158.

56. Miller, S. A. in Single-Cell Protein. R. I. Mateles and S. R. Tannenbaum, Eds., M.I.T. Press, Cambridge, Mass. (1968) pg. 79.

57. Waslien, C. I., Calloway, D. H., Morgen, S., Costa, F. J. Food Sci. (1970) 35, 294.

58. Labuza, T., Jones, K. J. Food Sci. (1973) 38, 177.

59. Labuza, T. P., Jones, K. A., Sinskey, A. J., Gomez, R., Wilson, S. and Miller, B. J. Food Sci. (1972) 37, 103.

60. Wimpenny, J. W. Process Biochem. (1967) 2, 41.

61. Vananuvat, P., Kinsella, J. E., J. Agr. Fd. Chem. (1975) 23, 216.

62. Dunhill, P., Lilly, M. in Single-Cell Protein II. S. R. Tannenbaum and D.I.C. Wang, Eds. M.I.T. Press, Cambridge, Mass. (1975) pg. 179.

63. Hedenskog, H., Mogren, H. Biotechn. Bioengn. (1973) 15, 129.

64. Carenburg, C., Haden, C. Biotechn. Bioeng. (1970) 12, 167.

65. Mogren, H., Lindbloom, M Biotechn. Bioeng. (1974) 16, 261.

66. Johnson, J. C. Yeasts for Food and Other Purposes, Noyes Data Corp., Park Ridge, N. J. (1977) p. 138.

67. Phaff, H. J. in Food Proteins. R. Feeney and J. Whitaker, Eds. Am. Chem. Soc., Washington, D.C. (1977) pg. 244.

68. Cunningham, S. D., Cater, C. M. and Mattil, K. F.   J. Food Sci. (1977) 40, 732.
69. Lindbloom, M.   Biotechnol. Bioeng. (1977) 19, 199.
70. Pringle, J.   Methods in Cell Biol. (1975) 12, 158.
71. Castro, A. C., Sinskey, A. J., Tannenbaum, S. R.   J. Appl. Microbiol. (1971) 22, 422.
72. Newell, J. A., Seeley, R. D., Robins, E. A.   U.S. Patent 3,867,255 (1975).
73. Lindbloom, M.   Biotechnol. Bioeng. (1974) 16, 495.
74. Vananuvat, P., Kinsella, J. E.   Biotechnol. and Bioeng. (1978) 20, 1329.
75. Cheftel, C. in Food Proteins.   J. Whitaker and S. Tannenbaum Eds., AVI Publ. Co., Westport, Conn. (1977) pg. 401.
76. Shetty, K. J. and Kinsella, J. E.   Biotechnol. Bioeng. (1978) 20, 755.
77. Tannenbaum, S.   Food Technol. (1971) 25, 962.
78. Litchfield, J. H.   J. Food Technol. (1977) 31, 175.
79. Akin, C. U. S. Patent 3,833,552 (1976).
80. Newell, J. A., Robins, E. A., Seeley, R. D., U.S. Patent 3,867,555 (1975).
81. Lipinsky, E. S., Litchfield, J. H.   Food Technol. (1974) 28, 16.
82. Robins, E. A., Sucker, R. W., Schuldt, E. H., Sidoh, D. R., Seeley, R. D., Newell, J. A.   U.S. Patent 3,887,431 (1975).
83. Lindbloom, M. Levensmit-Wiss Technol. (1976) 7, 285.
84. Vananuvat, P., Kinsella, J. E.   J. Agric. Food Chem. (1975) 23, 613.
85. Rha, C. in Single-Cell Protein II.   S. R. Tannenbaum and D.I.C. Wang, Eds., M.I.T. Press, Cambridge, Mass.   (1975) pg. 587.
86. Reed, G., Peppler, H. J.   Yeast Technology. AVI Publ. Co., Westport, Conn.   (1973).
87. Hermansson, A. M., Olsson, U., Holmberg, B.   Rebens-Wiss u Technol. (1974) 7, 176.
88. McElwain, M., Richardson, T., Amundson, C. H.   J. Milk Food Tech. (1975) 38, 521.
89. Pearce, K. N., Kinsella, J. E.   J. Agr. Food Chem. (1978) 26, 716.
90. Newell, J., Seeley, R., Robins, E. U.S. Patent 3,867,255 (1975).
91. Meyer, E. W. and Williams, L. D. in Food Proteins. R. Feeney and J. Whitaker, Eds. Am. Chem. Soc., Washington, D.C. (1977) pg. 52.

RECEIVED October 18, 1978.

Conformation and Functionality of Milk Proteins

C. V. MORR

Department of Food Science, Clemson University, Clemson, SC 29631

We are indebted to numerous researchers who have contributed
sufficient basic knowledge to enable us to unravel the complexi-
ties of the milk protein system, e.g., the caseins and whey
proteins and their subfractions. These findings have been collec-
ted, evaluated and organized into a comprehensive report by the
"Milk Protein Nomenclature Committee" (1). This latter report
contains excellent details on the primary structure and physico-
chemical properties of most of the casein monomer subunits and
the whey proteins, as well as for most of their genetic polymorphs. Milk proteins are widely recognized for their superior
nutritional, organoleptic, and functional properties, as components of milk and in milk-containing food formulations. Although
the caseinates have superior functional properties, especially
for stabilizing emulsions and foams, there is a need to improve
the functionality of the whey protein concentrates. For example,
the whey protein concentrates generally exhibit poor functional
properties in food applications requiring high solubility.

This paper draws heavily upon the "Nomenclature Committee
Report" (1) as well as several recent comprehensive reports that
have considered the primary structure and conformation of the
casein monomer subunits and how they are assembled into submicellar aggregates and casein micelles (2, 3). These basic relationships were utilized to develop additional projections relating to
the conformation and functional properties of the major milk proteins, e.g., commercial caseinates and whey protein concentrates
in food applications.

Fractionation and Distribution of Major Milk Proteins.

The fractionation scheme and distribution data for the major
milk proteins and their subfractions are given in Figure 1 (1).
Although the caseins are easily separated from whey proteins by
adjusting milk to pH 4.6-5.0, further separation and purification
of the individual caseins is extremely difficult, due to their
strong interaction, and requires the most sophisticated protein

0-8412-0478-0/79/47-092-065$05.00/0

fractionation techniques available. Similarly, the individual
whey protein components are separated and purified only by application of elaborate fractionation techniques.

Nomenclature and Physico-chemical Properties of the Caseins.

The four major casein fractions contain subfractions and genetic polymorphs (Table 1), each of which possesses a unique primary structure and associated physico-chemical properties ($\underline{1}$).
For example, α_s - casein contains one major subfraction (α_{s1}-Cn)
plus several minor components. The four genetic variants of α_{s1}-
casein have similar isoionic points, which would be expected from
their similar primary structures. The β-caseins, κ-caseins, and
"whole α-caseins" have also been fractionated and their subfractions characterized (Table 1). The isoionic points for all of
these casein subunits range from about 4.9 to 6.0 and their molecular weights range from about 11,500 to 24,000.

Primary Structure of the Major Casein Monomer Subunits.

The primary structures (Figures 2, 3 and 4) are given for
α_{s1}casein, β-casein, and κ-casein ($\underline{1}$). The most significant aspects of these primary structures are their disproportionate dis-
tribution of acidic amino acids, serine phosphate
groups and hydrophobic amino acids along the polypeptide chain.
As demonstrated by Bloomfield and Mead ($\underline{2}$), the uneven distribution of amino acids for the major casein monomer subunits leads to
highly charged (negative) regions, which are separated from the
strongly hydrophobic regions, along the polypeptide chains. These
amphiphilic molecules are highly susceptible to intermolecular interaction and polymerization through the formation of hydrophobic
and ionic, especially Ca bonds. Slattery ($\underline{3}$) used this and other
relevant information to develop conformational models for each of
the major casein monomer subunits. Inherent in these models is
the consideration that the uniform distribution of proline along
the polypeptide chain results in a random coil, with little helical structure. The model for α_s-casein that best fits the data
is a compact, prolate ellipsoid containing most of the hydrophobic amino acid residues, plus a 40 amino acid residue chain containing most of the acidic amino acids and serine phosphates that
extend into the aqueous phase as a loop ($\underline{3}$). The compact hydrophobic region is stabilized by intramolecular hydrophobic bonding,
whereas the exposed, acidic loop is readily accessible for interaction with adjacent casein molecules through formation of ionic,
mainly Ca, bonds. The model developed by Slattery ($\underline{3}$) for β-ca-
sein also contains a compact hydrophobic region, with about the
same dimensions as for α_s-casein, and with a strongly acidic
polypeptide chain, e.g., amino acid residues 1-25, exposed to the
aqueous media. The model for κ-casein($\underline{3}$) also contains a compact
hydrophobic region, as above, plus a freely exposed polypeptide
chain containing a high portion of acidic amino acids as well as

Table I. Amount and Selected Properties of Milk Proteins and Their Components[a]

Protein	Approx % of skimmilk protein	Number of components[b]	Isoionic point	Molecular weight[c]
Caseins				
α_s-Caseins	45-55	9	4.92-5.35	22,068-22,723
κ-Caseins	8-15	2	5.37	19,005-19,037
β-Caseins	23-35	7	5.20-5.85	23,939-24,089
γ-Caseins	3-7	9	5.80-6.0	11,556-20,629
Whey Proteins				
β-Lactoglobulins	7-12	4	5.35-5.41	18,275-18,362
Bovine serum albumin	.7-1.3	1	5.13	66,500-69,000
α-Lactalbumins	2-5	2	4.20-4.5	14,146-14,174
Immunoglobulins	1.9-3.3	4	5.50-8.3	150,000-1 million
Proteose peptones	2-6	4	3.30-3.7	4,100-40,800

[a]Adapted from Whitney et al., 1976

[b]Total of genetic variants and minor components

[c]Calculated from primary structure where possible

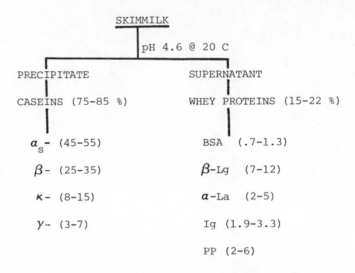

Figure 1. *Fractionation and distribution of the major milk proteins*

10   20
H.Arg- Pro- Lys–His–Pro–Ile – Lys–His– Gln–Gly–Leu–Pro–Gln–|Glu– Val –Leu–Asn–Glu–Asn–Leu–|
                                                            Absent in variant A

30   40
|Leu–Arg–Phe–Phe –Val–Ala|–Pro–Phe–Pro–Val– Phe–Gly –Lys–Glu–Lys–Val –Asn–Glu–Leu–

50   60
Ser – Lys–Asp–Ile – Gly–Ser – Glu–Ser –Thr–Glu–Asp–Gln –|Ala|– Met–Glu–Asp–Ile – Lys–Glu–Met–
              P   P   ThrP (variant D)

70   80
Glu–Ala –Glu–Ser –Ile – Ser – Ser – Ser –Glu–Glu–Ile – Val –Pro–Asn–Ser –Val –Glu –Gln–Lys –His –
              P   P   P   P   P

90   100
Ile – Gln–Lys–Glu–Asp –Val –Pro–Ser – Glu–Arg–Tyr– Leu–Gly –Tyr– Leu–Glu–Gln– Leu–Leu–Arg–

110   120
Leu –Lys–Lys –Tyr–Lys –Val –Pro–Gln–Leu–Glu–Ile – Val –Pro–Asn–Ser – Ala –Glu–Glu–Arg–Leu–
                                                             P

130   140
His – Ser – Met– Lys– Gln–Gly –Ile – His – Ala–Gln– Gln–Lys–Glu–Pro –Met–Ile – Gly–Val –Asn–Gln–

150   160
Glu – Leu–Ala –Tyr–Phe –Tyr–Pro–Glu–Leu–Phe –Arg –Gln–Phe–Tyr –Gln–Leu–Asp–Ala –Tyr–Pro–

170   180
Ser– Gly–Ala –Trp –Tyr–Tyr–Val –Pro– Leu–Gly–Thr–Gln –Tyr–Thr–Asp–Ala –Pro–Ser – Phe –Ser–

190   199
Asp–Ile – Pro– Asn–Pro–Ile – Gly–Ser –Glu–Asn–Ser –|Glu|– Lys–Thr –Thr–Met–Pro– Leu–Trp.OH
                                                     Gly (variant C)

Journal of Dairy Science

Figure 2. *Primary structure of bovine α_s-Cn-B (1)*

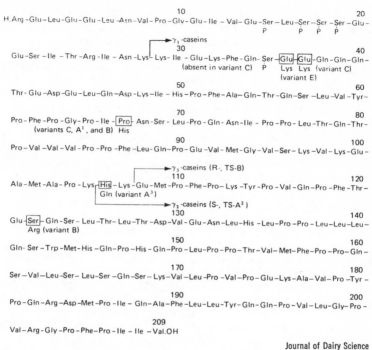

Figure 3.   Primary structure of bovine β-Cn-A² (1)

Figure 4.   Primary structure of bovine κ-Cn-B (1)

the negatively charged glycomacropeptide (GMP) group. It is believed that the exposed, GMP-containing polypeptide chain is
responsible for κ-casein's pronounced stabilizing action for the
milk casein micelle system, since cleavage of this entity by
enzymatic rennet modification completely removes its stabilizing
ability. Comparison of the distribution of the number of important amino acid types for each of the caseins (Table 2) provides
an indication of their overall capacity for aggregation and
interaction via hydrophobic, ionic and disulfide bond formation.
It will be noted that κ-casein is the only monomer subunit that
contains a disulfide group, and this is undoubtedly responsible
for its ability to self aggregate as well as to interact with β-
lactoglobulin during heat processing of milk.

Association of Casein Monomer Subunits.

All three of the casein monomer subunits exhibit a strong
capacity to interact at neutral pH. Addition of Ca ions
strongly promotes their tendency to aggregate, presumably by
cross-linking carboxyl and phosphate ester groups located in their
exposed, acidic peptide chains. Addition of CA probably also
promotes aggregation of the casein monomer subunits by reducing
their molecular charge. It is possible that, due to steric hin-
drance, κ-casein's negatively charged GMP group may be unable to
participate in intermolecular Ca cross-links with adjacent casein
monomers and aggregates. This may account for the micelle stabilizing function discussed above for κ-casein. These casein-casein
interactions, especially those involving β-casein, are strongly
temperature-dependent; a phenomenon that is important for asso-
ciation/dissociation of these monomer subunits with and from the
milk casein micelle system (4).

κ-casein also contains two ½ Cys residues per monomer subunit
and is thus capable of interacting with the whey proteins, e.g.,
mainly β-lactoglobulin, via the disulfide interchange mechanism
at temperatures at or above $65^{o}C$. This latter phenomenon is
believed to be important in providing colloidal stability to the
milk casein micelle system, as well as to the whey proteins, in
high temperature processed milk products. It has also been postulated that this latter interaction with β-lactoglobulin may
alter the availability of κ-casein in the micelle, and thus has a
detrimental effect upon the cheese making properties of milk (4).

The Casein Micelle System.

The major caseins exist in milk as highly structured, spherical aggregates, consisting of 450 to 10,000 subunits (3),
commonly referred to as micelles. The important physico-chemical
properties of the micelles are summarized in Table 3. Casein
micelles are synthesized in vivo by biochemically controlled
processes, which have not been totally characterized (5). Even

Table II. Amino Acid Composition of the Major Casein Monomers

CASEIN	TOTAL	ACIDIC[1]	SERINE P	BASIC[2]	PROLINE	½ CYS	STALIC ACID
α_{s1}-B	199	42	8	25	17	0	0
β-A[2]	209	25	5	20	35	0	0
κ-B	169	25	1	17	20	2	1

[1]Asp, Glu and Tyr
[2]Lys, Arg and His

Table III.   Physico-Chemical Properties of Casein Micelles in Milk[a]

Casein, micelle/total	
0-5 C	75-80 %
25 C	95-98 %
Diameter, nm	100-250
Sedimentation Constant, $S_{20, w}$	$8-22 \times 10^2$ S
Molecular Weight, Daltons	$2-18 \times 10^8$
Solvation, g water/g dry micelle	
0-5 C	2-3
25 C	1.6-2
Voluminosity, cc/g micelle	
25 C	3.5-6

[a]Adapted from Morr, 1975

though they are remarkably stable in milk under normal conditions, treatments such as cooling and warming (between 5 and 37°C), adjusting the pH by chemical or micro-biological means, adjusting Ca and other ion concentrations, heating, evaporation and drying, enzymatic modification (as with rennet) all effect the stability of the micelles (4). There have been a number of excellent papers published that provide an insight into the complexity of the milk casein micelle structure as well as an understanding of the factors that stabilize it. Scientists who have made special contributions in this regard are: Waugh, et al. (6), Payens (7), Garnier and Ribadeau-Dumas (8), Morr (9), Rose (10) and Bloomfield and Mead (2). Slattery (3) recently proposed a model for the micelles and their mechanism of formation, based upon the amphiphilic nature of the individual casein monomer subunits and their submicellar aggregates. Although Slattery's model considers the role of the submicellar aggregates of the three major caseins (α_s-, β-, and κ-casein) and their assembly mechanism, it ignores the importance and irreversible structure of "colloidal phosphate" (11) in the micelle.

Commercial Caseinates.

Casein is isolated from milk and produced as Na, K, and Ca caseinates by the general steps outlined in Figure 5. Both the acidification and neutralization steps profoundly affect the structure and assembly of resulting caseinate particles. These derived caseinate aggregate particles have little similarity to milk casein micelles. Acidification to isoelectric conditions, e.g., pH 4.5-5.0, completely dissipates the colloidal phosphate structure and frees the casein precipitate of Ca and other inorganic ions. Neutralization of the casein precipitate resolubilizes the casein, presumably by altering its charge sufficiently to overcome intermolecular hydrophobic bonding. The resulting caseinate system contains aggregates of the monomer subunits in a range of sizes, depending upon temperature and electrolyte composition (12). The orientation of the individual casein monomer subunits within these aggregates is unknown, but it may be assumed that, due to the rapidity of the acidification/resolubilization reactions, they are randomly associated. Based upon the strong sensitivity of these aggregates to pH and ionic adjustments, as well as to enzymatic modification, it is likely that the subunits are arranged with an orientation that provides their hydrophilic, acidic-polypeptide-chain sequence maximum contact with the aqueous medium. Commercial caseinates, prepared as Na and K caseinates exhibit improved solubility and functionality compared to Ca caseinate. This latter phenomenon is probably due to larger sized and more strongly interacting Ca caseinate aggregates due to cross-linking by the divalent cations.

Comparison of the above commercial caseinates with milk casein micelles indicates that, under normal food processing condi-

tions, e.g., pH 6 to 7, caseinate aggregates are smaller and more
sensitive to pH and ionic fluctuation than are the colloidal
phosphate-containing milk micelles. For example, the milk micelle
system stabilizes a higher level of Ca ions than can the commercial caseinates, due to its incorporation into the relatively
nonreactive colloidal phosphate polymer structure (10). Although
there has been much conjecture on the subject, it is considered
highly doubtful that micelles resembling those in milk in every
detail, can be reformed by combining their individual components.
If this latter function could be accomplished, it would probably
permit the development of new caseinate forms with modified functional properties.

Functional Properties of Commercial Caseinates.

Although commercial caseinates are used in a wide variety of
applications within the food industry (13, 14), they are especially useful in those that utilize their excellent surfactant properties. The probable explanation for the excellent surfactant
properties of the caseinates lies in their unique amphiphilic conformation as well as their ordered aggregate structures, as described above. It has been reported that casein micelles rapidly
associate on the surface of freshly-formed fat globules (15) to
stabilize them against coalescence and separation. Thus, it may
be postulated that caseinate, either as monomer subunits or their
aggregates, also rapidly migrate to and associate on freshly
formed air/water or oil/water interfaces of foams or emulsions to
stabilize them against collapse or coalescence. It is further
proposed that such emulsions and foams, due to the high availability of caseinate hydrophilic, acidic peptide chains to the aqueous phase, would exhibit a strong susceptibility to pH and Ca ion
fluctuations. Also, additional emulsifiers, foaming agents and
food ingredients that alter the availability and reactivity of the
exposed casein acidic chains would likely have a profound influence
upon the functional properties of the caseinates in foam and emulsion applications.

Enzymatic Modification of Casein.

The action of rennet upon casein micelles in milk illustrates
the tremendous importance of enzymatic modification upon the conformation, physico-chemical and functional properties of caseinates. Rennet specifically hydrolyzes the 105-106 linkage of the
κ-casein polypeptide chain which, although located within the
micelle, is highly accessible to the enzyme. The effect of this
reaction is to release the strongly acidic GMP from κ-casein
monomer subunits, thereby reducing the magnitude of the negative
charge on the micelle sufficiently to permit them to associate
into a gel structure (4).

Properties of the Whey Proteins and Their Subfractions.

As with the caseins, whey proteins have been further fractionated and these subfractions have been characterized (1). The
primary structures of two of these proteins, β-lactoglobulin and
α-lactalbumin are given in Figures 6 and 7. Comparison of these
primary structures with those of the caseins indicates several major differences which account for their distinct physico-chemical
and functional properties. In contrast to the caseins, whey proteins exhibit a rather uniform distribution of acidic/basic and
hydrophobic/hydrophilic amino acids along their polypeptide
chains. Thus, they lack the amphiphilic nature of the casein monomer subunits, but are rather present in a compact, globular
conformation (16), The substantially lower proline content in the
whey protein molecules also permits a globular conformation with
a substantial helical content, which explains their strong susceptibility to denaturation by heat and similar treatments (4). Isoionic points of the major whey proteins range from 4.2 to 5.4
(Table 1), which are comparable to those of the caseins.

Denaturation of whey proteins has been shown to unfold their
polypeptide chains and convert their conformation from a globular
to an extended form that facilitates their self interaction (aggregation) and interaction with κ-casein by disulfide interchange,
ionic and hydrophobic bonding (4).

Although β-lactoglobulin exists as a monomer with a molecular
weight of 18,000 at pH's below 3.5, it associates to form an
octamer with molecular weight of 144,000 at pH's in the 3.7 to
5.1 range and exists as a dimer with a molecular weight of 36,000
at pH 5.1 (16). These pH-dependent interactions are due to changes in ionization of acidic and basic amino acids which have an
effect upon the formation and dissipation of hydrophobic and
disulfide bonds.

Although β-lactoglobulin and α-lactalbumin are capable of the
above association/dissociation reactions, they probably have little
importance in determining functional properties in food applications in the pH 6 to 7 range. However, protein concentration, pH
and other related factors do affect their susceptibility to heat
denaturation (17).

Preparation of Whey Protein Concentrates.

In contrast to the caseins, whey proteins retain their solubility in the pH 4.5-5.0 range, provided they have not been
denatured. It is therefore relatively difficult to recover and
purify undenatured protein concentrates on a commercial scale.
Processes that separate the whey proteins from the low molecular
weight, nonprotein components of whey have been used with only
moderate success to date (18). Such processes utilize ultrafil-
tration/reverse osmosis membrane technology, gel filtration by the
basket centrifuge technique, polyvalent ion precipitating agents

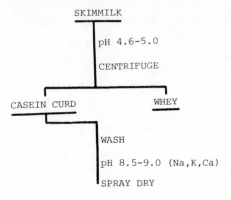

SKIMMILK

pH 4.6-5.0

CENTRIFUGE

CASEIN CURD   WHEY

WASH

pH 8.5-9.0 (Na,K,Ca)

Figure 5. Procedure for preparation of
commercial caseinate

SPRAY DRY

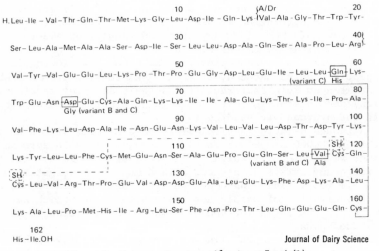

```
                           10              (A/Dr              20
H.Leu-Ile - Val-Thr-Gln-Thr-Met-Lys-Gly-Leu-Asp-Ile - Gln-Lys {Val-Ala - Gly-Thr-Trp-Tyr-
                           30                                40(
Ser-Leu-Ala-Met-Ala-Ala-Ser-Asp-Ile - Ser-Leu-Leu-Asp-Ala-Gln-Ser-Ala-Pro-Leu-Arg}
                           50                                60
Val-Tyr-Val-Glu-Glu-Leu-Lys-Pro -Thr-Pro-Glu-Gly-Asp-Leu-Glu-Ile - Leu-Leu-[Gln]-Lys-
                                                              (variant C)  His
                           70                                80
Trp-Glu-Asn-[Asp]-Glu-Cys-Ala-Gln-Lys-Lys-Ile - Ile - Ala-Glu-Lys-Thr-Lys-Ile - Pro-Ala-
              Gly (variant B and C)
                           90                                100
Val-Phe -Lys-Leu-Asp-Ala -Ile -Asn-Glu-Asn-Lys-Val-Leu-Val-Leu-Asp-Thr-Asp-Tyr -Lys-
                           110                               SH  120
Lys -Tyr-Leu-Leu-Phe -Cys-Met-Glu-Asn-Ser-Ala-Glu-Pro-Glu-Gln-Ser-Leu-[Val]-Cys-Gln-
                                                              (variant B and C)  Ala
SH                         130                               140
Cys-Leu-Val -Arg-Thr-Pro-Glu-Val-Asp-Asp-Glu-Ala-Leu-Glu-Lys-Phe-Asp-Lys-Ala-Leu-
                           150                               160
Lys-Ala-Leu-Pro-Met-His-Ile - Arg-Leu-Ser-Phe-Asn-Pro-Thr-Leu-Gln-Glu-Glu-Gln-Cys-
```

162
His-Ile.OH   **Journal of Dairy Science**

Figure 6. Primary structure of bovine β-Lg-A (1)

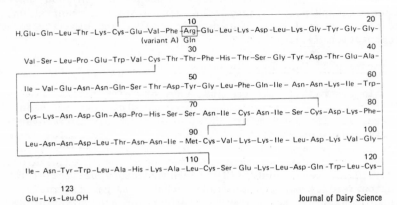

```
                        10                                   20
H.Glu-Gln-Leu-Thr-Lys-Cys-Glu-Val-Phe-[Arg]-Glu-Leu-Lys-Asp-Leu-Lys-Gly-Tyr-Gly-Gly-
                             (variant A)  Gln
                        30                                   40
Val-Ser-Leu-Pro-Glu-Trp-Val-Cys-Thr-Thr-Phe-His-Thr-Ser-Gly-Tyr-Asp-Thr-Glu-Ala-
                        50                                   60
Ile - Val-Glu-Asn-Asn-Gln-Ser-Thr-Asp-Tyr-Gly-Leu-Phe-Gln-Ile - Asn-Asn-Lys-Ile -Trp-
                        70                                   80
Cys-Lys-Asn-Asp-Gln-Asp-Pro-His-Ser-Ser-Asn-Ile - Cys-Asn-Ile - Ser-Cys-Asp-Lys-Phe-
                        90                                   100
Leu-Asn-Asn-Asp-Leu-Thr-Asn-Asn-Ile - Met-Cys-Val-Lys-Lys-Ile - Leu-Asp-Lys-Val-Gly-
                        110                                  120
Ile - Asn-Tyr-Trp-Leu-Ala-His-Lys-Ala-Leu-Cys-Ser-Glu-Lys-Leu-Asp-Gln-Trp-Leu-Cys-
```

123
Glu-Lys-Leu.OH   **Journal of Dairy Science**

Figure 7. Primary structure of α-La-B (1)

(carboxymethyl cellulose, hexametaphosphate, acrylic acid, etc.),
electrodialysis, and ion-exchange resin treatments to fractionate
and recover the whey protein concentrates which commonly contain
only 35 to 50% protein on a dry weight basis. Inherent in each
of these processes is the necessity to concentrate the dilute whey
protein solution under conditions that minimize denaturation of
the heat-sensitive whey proteins. Both of the commonly used concentration processes, e.g., vacuum evaporation and spray drying,
introduce a small but definite amount of protein denaturation into
the overall process (19).

Whey protein concentrate preparation processes based upon
heat denaturation to insolubilize and precipitate them from whey
have been developed (20). Such whey protein concentrate products
are suitable for applications in pasta products (21) due to their
insolubility and lack of stickiness.

Functional Properties of Whey Protein Concentrates.

Although whey protein concentrates possess excellent nutritional and organoleptic properties, they often exhibit only partial solubility and do not function as well as the caseinates for
stabilizing aqueous foams and emulsions (19). A number of compo-
sitional and processing factors are involved which alter the
ability of whey protein concentrates to function in such food
formulations. These include: pH, redox potential, Ca concentration, heat denaturation, enzymatic modification, residual poly-
phosphate or other polyvalent ion precipitating agents, residual
milk lipids/phospholipids and chemical emulsifiers (22).

It is likely that the inability of whey proteins to function
as well as caseinate in stabilizing foams and emulsions is due to
conformational and structional differences in the two proteins.
It is therefore postulated that whey proteins, which lack an
amphiphilic conformation, do not orient sufficiently well at air/
water interfaces to stabilize foam or emulsion systems as effectively as caseinates.

Denaturation of the whey protein molecule, if produced at the
proper stage of the protein concentrate isolation/utilization process, can improve the functionality. The improvement in functionality is probably due to an unfolding of the molecule to expose
hydrophobic amino acid residues, thus making the protein more
amphiphilic and capable of orienting at the air/water or oil/water
interface. However, if the protein concentrate is denatured
during preliminary preparation stages, e.g., prior to or during
drying, their solubility and related functionality are adversely
affected.

Summary.

The conformation and physico-chemical properties of the major
casein and whey protein components, and their subfractions, are

related to the functional properties of commercial caseinate and whey protein concentrates used in the food processing industry. The effects of pH, temperature, ionic environment, heat denaturation, enzymatic modification and processing are also considered in this regard.

The major casein monomer subunits have random coil conformation that facilitates strong protein-protein interaction via hydrophobic and ionic bonding. The unique amphiphilic structure, which arises from separately clustered hydrophobic and negatively charged (acidic and ester phosphate) amino acid residues along the polypeptide chain, makes them susceptible to pH and Ca ion concentration effects. This amphiphilic nature is probably responsible for the excellent surfactant properties of commercial caseinate in a variety of food applications.

The major whey proteins have compact, globular conformations, with a substantial amount of helical structure. This conformation is probably due to the rather uniform distribution of acidic/basic and hycrophilic/hydrophobic amino acids along their polypeptide chains. Their sensitivity to heat and other denaturing agents is due to the high content of sulfhydryl/disulfide groups, which stabilize their derived, random coil conformation by the disulfide bond interchange mechanism. Exposure of sulfhydryl, hydrophobic and acidic amino acid residues during denaturation makes the whey proteins susceptible to self polymerization and interaction with casein, e.g., mainly κ-casein. Although heat denaturation generally reduces the solubility and functionality of whey proteins, it can be utilized, if conducted at the proper point in the process, to improve their functionality.

Literature Cited

1 Whitney, R. McL., Brunner, J. R., Ebner, K. E., Farrell, H. M., Jr., Josephson, R. V., Morr, C. V., Swaisgood, H. E., J. Dairy Sci. (1976) 59, 795.
2 Bloomfield, V. A., Mead, R. J., Jr., J. Dairy Sci. (1975) 58, 795.
3 Slattery, C. W., J. Dairy Sci. (1976) 59, 1547.
4 Morr, C. V., J. Dairy Sci. (1975) 58, 977.
5 Farrell, H. M., Jr., J. Dairy Sci. (1973) 56, 1195.
6 Waugh, D. F., Creamer, L. K., Slattery, C. W., Dresdner, G. W., Biochem. (1970) 9, 786.
7 Payens, T. A. J., J. Dairy Sci. (1966) 49, 1317.
8 Garnier, J., Ribadeau-Dumas, B., J. Dairy Res. (1970) 37, 493.
9 Morr, C. V., J. Dairy Sci. (1967) 50, 144.
10 Rose, D., J. Dairy Sci. (1965) 48, 139.
11 Rose, D., Dairy Sci. Abstr. (1969) 31, 171.
12 von Hippel, P. H., Waugh, D. F., J. Am. Chem. Soc. (1955) 77, 4311.
13 Muller, L. L., Dairy Sci. Abstr. (1971) 33, 659.
14 Borst, J. R., Food Technol. in Australia (1971) 23, 544.
15 Ogden, L. V., Walstra, P., Morris, H. A., J. Dairy Sci. (1976) 59, 1727.
16 Timasheff, S. N., "Symposium on Foods: Proteins and Their Reactions," 179-208, AVI Pub. Co., Inc., Westport, Conn., 1964.
17 Nielsen, M. A., Coulter, S. T., Morr, C. V., Rosenau, J. R., J. Dairy Sci. (1973) 56, 76.
18 Morr, C. V., Food Technol. (1976) 30, 18.
19 Morr, C. V., Swenson, P. E., Richter, R. L., J. Food Sci. (1973) 38, 324.
20 Panzer, C. C., Schoppet, E. F., Sinnamon, H. I., Aceto, N. C., J. Food Sci. (1976) 41, 1293.
21 Seibles, T. S., Cereal Foods World (1975) 20, 487.
22 Richert, S. H., Morr, C. V., Cooney, C. M., J. Food Sci. (1974) 39, 42.

RECEIVED October 23, 1978.

Aggregation and Denaturation Involved in Gel Formation

A.-M. HERMANSSON

SIK—The Swedish Food Institute, Fack, S-400 23 Göteborg, Sweden

When a protein dispersion is heated a gel may form. The gelation
mechanism, the gel structure and the gel properties are highly
dependant on processing conditions as well as factors such as
pH and ionic strength. Differences in gel structures are demonstrated in <u>Figures 1 and 2</u>, where electromicrographs of two
gel structures of exactly the same protein preparation, formed
at different ionic strengths are shown.

The aim of this lecture is to impart some insight into the mechanisms involved in gel formation. These mechanisms are determined
by the balance between forces underlying chain-chain and chain-
-solvent interactions. Mechanisms and conformations favored by
either of these interactions are listed in Table I.

Table I. Mechanisms and conformations

Chain-Solvent	Chain-Chain
mechanism(s)	mechanism(s)
solubilization	precipitation
dissociation	association
swelling	aggregation
denaturation	flocculation
	coagulation
conformation(s)	conformation(s)
coil	helix
	native structure
	three dimensional structure

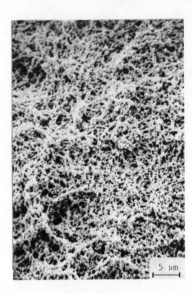

*Figure 1. Scanning electron micrograph of a 10% WPC gel in distilled water.
Chemical fixation was made with glutaraldehyde followed by critical-point drying.*

*Figure 2. Scanning electron micrograph of a 10% WPC gel in 0.2M NaCl.
Chemical fixation was made with glutaraldehyde followed by critical-point drying.*

The formation of gel networks as well as general viscosity
changes may involve mechanisms depending on chain-solvent as
well as chain-chain reactions.

Before going into the details of any reaction some definitions of
the terms on the right hand side need to be made. There is some
confusion in the therminology used, especially about the terms
association, aggregation, coagulation, and flocculation. A
more detailed discussion of these terms is given elsewhere (1).

Association normally refers to changes on the molecular level
such as monomer ⇌ dimer reactions, subunit equilibrium etc.
These reactions are characterized by weak bonds at specific
binding sites.

Contrary to "association" the terms "aggregation", "coagulation",
and "flocculation" refer to unspecified protein-protein interactions and the formation of complexes with higher molecular
weights.

Aggregation is a general term and will be used in this presentation as a collective term for protein-protein interactions.

Flocculation is an entirely colloidal phenomena where the interaction between protein molecules is determined by the balance
between electrostatic repulsion due to the electric double layer
and van der Waals' attraction (2).

Coagulation will be used for random aggregation which includes
denaturation of protein molecules.

Gelation is often an aggregation of denatured molecules. Contrary to coagulation where the aggregation is random, gelation
involves the formation of a continous network, which exhibits
a certain degree of order. The kinetics of the mechanisms *e.g.*
dissociation, swelling, denaturation, aggregation will determine the structure and the properties of the gel.

Denaturation and aggregation caused by heat treatment will be
discussed for two completely different protein systems namely
a soy and a whey protein system. Although the two protein
systems are different in character they both have the ability
to form a gel after heat treatment.

The major soy proteins, the 7S and the 11S globulins are characterized by complex quaternary structures easily undergoing as-
sociation-dissociation reactions. Salt has a unique stabilizing
effect on the quaternary structure of both the 7S and the 11S
globulins, which influences most physical properties of soy proteins.

The major whey proteins β-lactoglobulin and α-lactalbumin do not
have the same complex quaternary structure as the soy proteins.
The β-lactoglobulin normally exists in a monomer or dimer form
although one variant is known to exist in an octamer form at pH
4.65. Little or no association of α-lactalbumin has been observed
on the alkaline side of the isoelectric region. On the acid side
(pH 2-4) α-lactalbumin associates easily and the association has
previously been described as a "denaturation-like" process (3).

MATERIALS AND METHODS

Soy protein isolate. A soy protein isolate produced under mild
conditions on pilot plant scale was kindly provided by Central
Soya. The preparation procedure has been described elsewhere (1).
Protein content (N x 6.25) was 99.4% (dry weight).

Whey protein concentrate. The whey protein used was prepared by
ultrafiltration and spray drying. Protein content (N x 6.55)
was 68% (dry weight). Lipid content was 7.1% (dry weight). In
order to study heat induced aggregation by spectrophotometric
methods the turbidity of the dilute protein dispersions was too
high. The turbidity of whey protein dispersions is caused by lipids associated with proteins probably in the form of emulsified
oil droplets. This fraction was removed by precipitation at pH
4.5 from dispersions made in dist. water and separated by centrifugation at 40 000 xg.

Denaturation and aggregation studies. The methods used have been
described elsewhere (1).

Microstructure studies. Pieces of gels were fixed in 6% glutaraldehyde in dist. water or 0.2 M NaCl depending on the gel for
17 hrs. Dehydration was made in three steps: 70% ethanol, 95%
ethanol and abs. ethanol. The time for each step was 30 min.
The ethanol was exchanged by amylacetate by a series of amylace-
tate-ethanol mixtures: 50% amylacetate for 30 min; 70% amylacetate for one hour, and 100% amylacetate for two hours. The samples were critical-point-dried and coated with gold. The scanning
electron micrographs were taken in a Cambridge Stereoscan S4.

RESULTS AND DISCUSSION

Denaturation

The first reaction to be discussed is denaturation. Denaturation is involved in most structure forming processes although
gels may form from already denatured proteins. It is important
to control the denaturation process in order to obtain structures with the desired textural properties. Unfortunately most
studies on protein denaturation have been made by biochemists
mainly interested in the native structure. One definition given

by Tanford (4) is thus "Denaturation involves conformational changes from the native structure without alteration of the amino acid sequence". Basic studies are furthermore made in dilute solutions in order to avoid interactions between molecules. Such interactions are, however, fundamental for the formation of a three dimensional network.

When protein dispersions of soy and whey proteins are heated, they become turbid due to aggregation. Spectrophotometric methods can therefore not be used and few denaturation studies have been made on such systems. Differential scanning calorimetry is an interesting alternative method, since the physical state of the system is unimportant for this analytical tool. In this study a Perkin-Elmer DSC-2 was used and the effect of pH and NaCl concentration on the two protein systems was studied. Calorimetric studies of proteins make it possible to distinguish between the aggregation and denaturation involved in the structure formation of proteins. The effects of aggregation are considered negligible for the qualitative interpretation of DSC thermograms.

Figure 3 shows DSC thermograms of 10% soy protein dispersions at varying pH in distilled water. Two peaks can be observed in the pH range 4-9. The 7S globulin is responsible for the first and the 11S globulin for the second peak (1).

The denaturation process starts at the point where the curves begin to deviate from the baseline. This temperature is, however, difficult to identify in a reproducable way; especially for the second peak, since it is not clear whether the peaks overlap or not. Instead the intercept of the extrapolated slope of the peak and the baseline, was taken as a measure of the denaturation temperature (T_d). The temperature at the peak maximum (T_{max}) was also used for relative comparisons. From Figure 3 it can be seen that the highest denaturation temperature was obtained at pH 5 close to the isoelectric point. Only one peak was observed at low (2-3) and high (10) pH. It is obvious that the protein system is partially denatured at these pH's due to the high net charge favoring chain-solvent interactions.

The pH-dependence is much smaller in the presence of salt than in distilled water. Figure 4 shows DSC thermograms of 10% soy proteins at varying pH in 0.2 M NaCl.

The presence of salt seems to stabilize the protein structure against denaturation. The effect of salt on the denaturation temperature is more pronounced at pH values outside the isoelectric region. Table II gives an example of the effect of NaCl concentration on the denaturation temperature. It can be seen that the increase in T_d in the direction from 0 to 1.0 M NaCl is >20°C. This drastic effect of NaCl on the denaturation tempe-

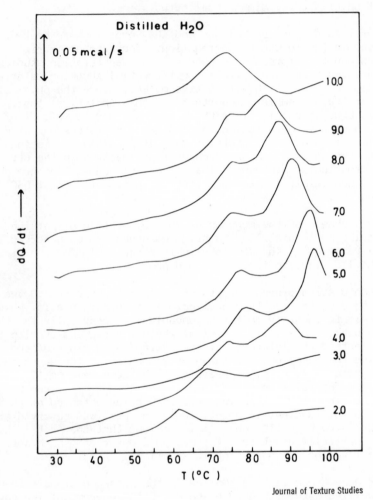

Figure 3.   DSC thermograms of 10% soy protein dispersions in distilled water at pH 2.0–10.0. The sensitivity was 0.5 mcal sec^{-1} and the heating rate 10°C min^{-1} (1).

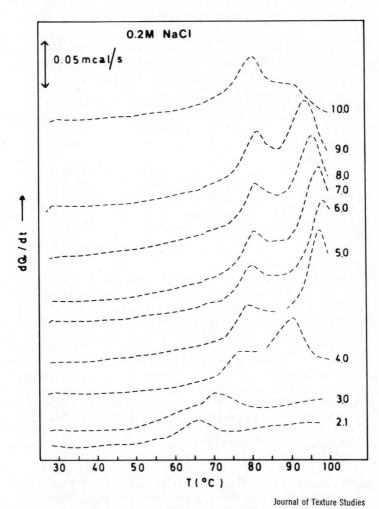

Figure 4. DSC thermograms of 10% soy protein dispersions in 0.2M NaCl at pH 2.0–10.0. The sensitivity was 0.5 mcal sec^{-1} and the heating rate 10°C min^{-1} (1).

rature is due to its effect on the quaternary structure of soy
proteins. Salt stabilizes the quaternary structure against dissociation and denaturation (5).

Table II. Endothermic DSC peak characteristics as a function
of NaCl concentration at pH 7.0

NaCl conc. (M)	Peak 1		Peak 2	
	T_d (°C)	T_{max} (°C)	T_d (°C)	T_{max} (°C)
0	67.0	76.0	80.0	91.0
0.01	68.0	76.5	81.0	91.0
0.05	70.0	77.0	84.5	92.5
0.10	72.0	78.5	86.0	94.0
0.25	74.0	80.5	90.0	97.0
0.50	78.5	85.0	97.0	101.0
1.0	87.0	92.0	103.0	103.0
2.0	97.0	102.0	>103.0	>113.0

(Hermansson, 1977 a)

The major whey proteins β-lactoglobulin and α-lactalbumin do not
have the same complex quaternary structure and a similar stabilizing effect of NaCl was not found when dispersions of whey
proteins at various pH were studied in distilled water and in
0.2 M NaCl.

Figure 5 shows DSC thermograms of 10% dispersions based on the
protein content of the whey protein concentrate at varying pH in
distilled water. Two peaks can be observed, one small at pH >4.0
and one bigger at pH 2-9. The small peak probably corresponds to
α-lactalbumin and the big peak to β-lactoglobulin. That the smaller peak disappeared at pH <4.0 is in agreement with the previously reported data on conformations changes of α-lactalbumin on the
acid side of the isoelectric region.

The highest denaturation temperature of the second peak is found
at pH 4.0 - 4.5. It is further seen that the second peak is
bigger and more well defined at pH <4.5. The area under the
peaks is proportional to enthalpy involved in the denaturation
process.

Even if salt had little influence on the position of the peak
onsets, salt seemed to have an effect on the areas under the
peaks for whey proteins. Figure 6 shows DSC thermograms of 10%
whey protein dispersions at varying pH in 0.2 M NaCl.

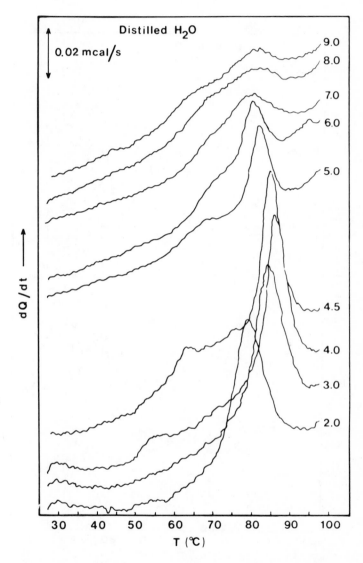

Figure 5. DSC thermograms of 10% whey protein dispersions in distilled water at pH 2.0–9.0. The sensitivity was 0.2 mcal sec⁻¹ and the heating rate was 10°C min⁻¹.

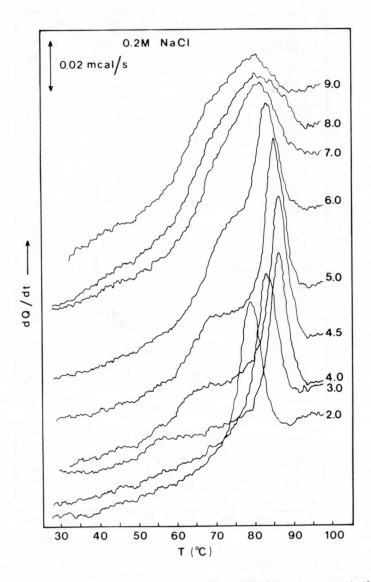

Figure 6. DSC thermograms of 10% whey protein dispersions in 0.2M NaCl at pH 2.0–9.0. The sensitivity was 0.2 mcal sec⁻¹ and the heating rate was 10°C min⁻¹.

In order to calculate enthalpies from this type of data much more work has to be done.

From both the DSC-thermograms of whey and soy proteins it can be concluded that the native structure is most stable against denaturation in the isoelectric region where the net charge is low.

Aggregation

Apart from denaturation where chain-solvent interactions are favored, heating may cause aggregation due to protein-protein interactions. Turbidity measurements may be used to follow aggregation reactions. They are easy to perform but it is not possible to say whether an increase in turbidity is due to changes in the number, the size or the optical properties of the particles.

Figure 7 shows the aggregation of whey proteins in distilled water at varying pH. It is seen that conditions favoring denaturation such as high and low pH had the opposite effect on aggregation. The tendency for aggregation was strongest in the isoelectric region.

The presence of salt had a positive effect on aggregation. The aggregation at varying pH in 0.2 M NaCl is shown in Figures 8 and 9.

The salt dependency is also illustrated in Figure 10 which shows the effect of NaCl concentration on the aggregation of whey proteins at pH 7.

The presence of salt promotes aggregation due to reduction of the diffuse part of the electric double layer. According to a review by Franks and Eagland, 1975 (6) salt effects on the electric double layer are to be found at $I \lesssim 0.1$ for univalent ions.

The aggregation behaviour of soy proteins was somewhat different from that of whey protein. This is illustrated in Figures 11 and 12, which show the aggregation of 0.5% soy protein dispersions in 0.2 M NaCl at varying pH. Similar to the behaviour of whey protein, aggregation was suppressed at high and low pH and no aggregation was observed at $3.0 \lesssim pH \gtrsim 11.0$. This is due to intermolecular repulsion forces at the high net charge, protein-solvent interactions being favored rather than protein-protein interactions. Apart from thermal aggregation, a reversible type of aggregation occurred for the soy protein system at intermediate pH. This type of aggregation was favored by lower temperatures. All measurements were made after cooling to $25^{\circ}C$.

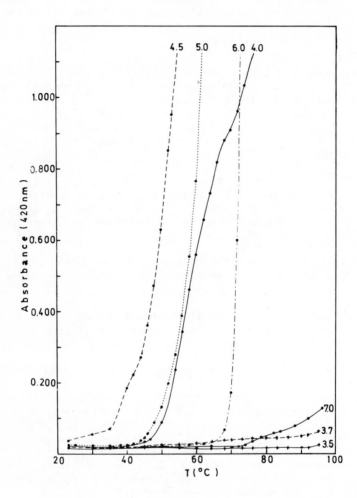

*Figure 7.   Turbidity as a function of heating temperature of 0.9% whey protein
dispersions in distilled water at pH 3.5–7.0*

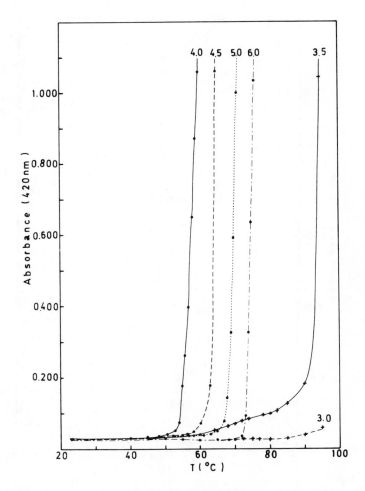

Figure 8. Turbidity as a function of heating temperature of 0.9% whey protein dispersions in 0.2M NaCl at pH 3.0–6.0

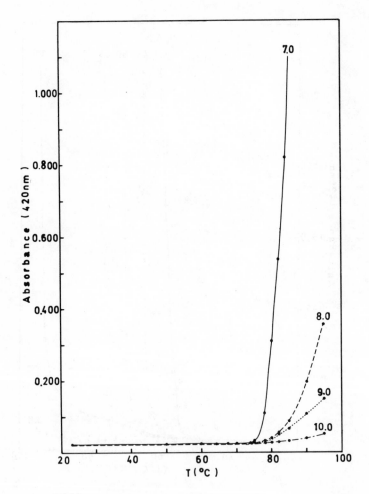

Figure 9.   *Turbidity as a function of heating temperature of 0.9% whey protein dispersions in 0.2M NaCl at pH 7.0–10.0*

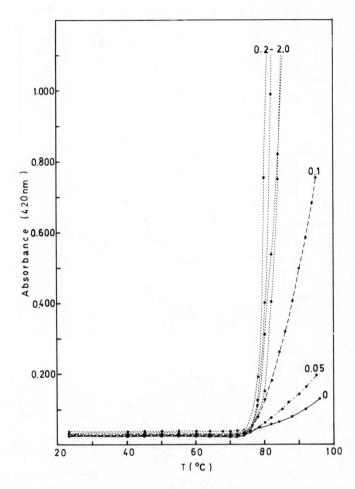

Figure 10. Turbidity as a function of heating temperature of 0.9% whey protein dispersions at pH 7.0 in 0–2.0M NaCl

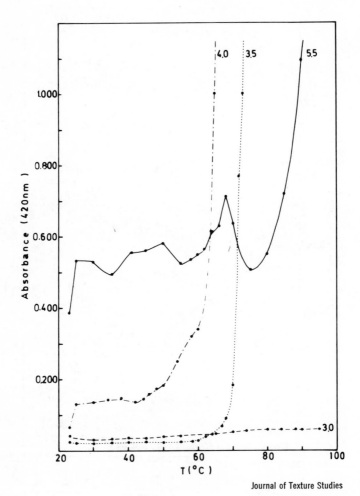

*Figure 11.   Turbidity as a function of heating temperature of 0.5% soy protein
dispersions in 0.2M NaCl at pH 3.0–5.5 (1)*

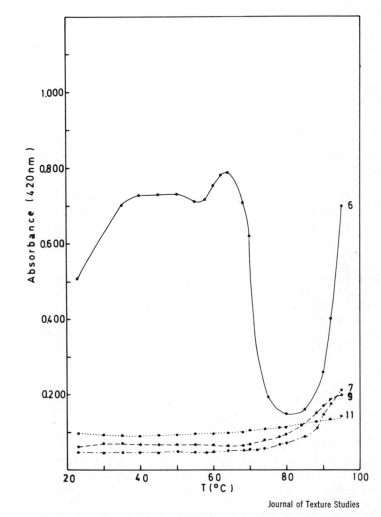

Figure 12. Turbidity as a function of heating temperature of 0.5% soy protein dispersions in 0.2M NaCl at pH 6.0–9.0 (1)

The formation of reversible aggregates has previously been observed by Catsimpoolas *et al.*, (7) with the glycinin or the 11S
fraction. They suggested that the reversible aggregation was
due to intermolecular interactions of undissociated molecules
and that aggregates formed at temperatures above 70°C originated
from dissociated subunits.

The soy protein isolate contains several proteins but additional
experiments indicated, that the glycinin and the 11S fraction
alone accounted for the reversible aggregation (1).

Salt was found to have a very special effect on the aggregation
of soy protein as can be seen from Figure 13.

At 1% concentration and pH 7.0 the aggregation was increased by
NaCl concentration to 0.2 M NaCl, whereafter it was decreased
by further increases in NaCl concentration. Reversible aggregation was formed at 0.1 M NaCl.

Both association and aggregation thus seem to be favored in
0.1 - 0.2 M NaCl. At NaCl concentrations below 0.1 M
dissociation into subunits may occur, but aggregation is suppressed due to the electric double layer and repulsion forces
in the absence of mobile counter ions.

At higher ionic strength, NaCl stabilizes against aggregation.
In this concentration range, salt has a unique effect on soy
proteins. It may stabilize them against aggregation, as well as
against denaturation. At first sight this is unexpected, since
it appears unlikely that protein-protein and protein-solvent
interactions should be favored by the same agent. An explanation
can be given if the cause is the stabilization of the quaternary
structure. This means that salt favors protein-protein interactions during heat treatment by stabilizing against dissociation of the quaternary structure and thereby protects against
aggregation as well as denaturation.

Three dimensional network
The results shown on aggregation and denaturation give some insight into structure forming processes of globular proteins.
Conditions favoring denaturation like high and low pH had the
opposite effect on aggregation. The presence of salt promoted
aggregation of whey proteins. Due to its unique stabilizing
effect on the quaternary structure of soy proteins, salt had a
suppressing effect on aggregation as well as denaturation.

It has already been said that the kinetics of the denaturation
and aggregation reactions are important for the final properties of the heat induced gel structure. Apart from environ-

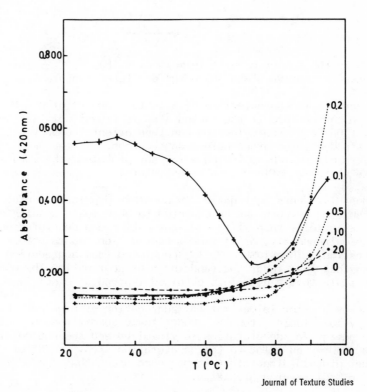

Figure 13.   *Turbidity as a function of heating temperature of 1% soy protein dispersions at pH 7.0 in 0–2.0M NaCl (1)*

mental conditions the heat gradient may influence aggregation.
This is illustrated in Figure 14, which shows the turbidity
as a function of heating temperature for a whey protein dispersion
at two different heat gradients.

In recent years very little work has been published on the kinetics of gel formation. In 1948 Ferry (8) suggested the following
model

$$xP_n \xrightarrow{\text{(1)}} xP_d \underset{(\longleftarrow)}{\overset{\text{(2)}}{\longrightarrow}} (P_d)_x$$

where x is the number of protein molecules (P).
n denotes the native state and d the denatured state.

A gel network is characterized by a certain degree of order.
This can be obtained if the second step is slower relative the
first. The denatured molecules can then orient themselves to
a certain degree of order before aggregation. If the second step
is reversible, which is the case for soy proteins, the intermediate state is defined as a progel state.

In a mechanism where aggregation is suppressed prior to unfolding,
the resulting network can be expected to show lower opacity and
higher elasticity than if random aggregation and denaturation
occur simultaneously, or if random aggregation occurs before
denaturation. Thus Tombs (9, 10) concluded from aggregation
studies that the higher the randomness of aggregation the more
likely it is that a coagel is obtained instead of a gel.

The energy barrier is maximal for aggregation at high net charge
and very low ionic strength. Under these conditions the activation energy for denaturation is minimized and the aggregation
can be suppressed prior to the denaturation step. Both for soy
and whey protein dispersions it has been shown that the addition
of 0.2 M NaCl favor aggregation.

Apart from studies on denaturation and aggregation, the final
gel structures have been investigated by electron microscopy.
An example is given by Figures 1 and 2 which show SEM micrographs of whey protein gels in the absence and presence of
0.2 M NaCl.

The differences between the gel structures are striking and the
addition of salt gave rise to a coarse aggregated gel structure.
Similar salt effects have been observed for other protein gels
such as soy protein and serum albumin gels.

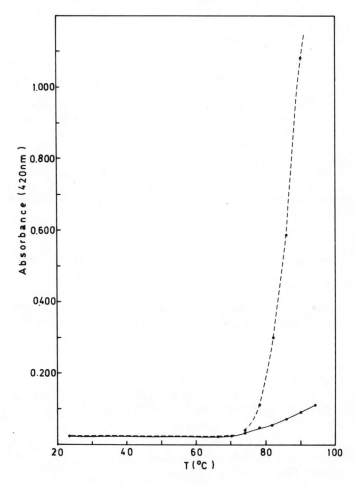

Figure 14.   Turbidity as a function of heating temperature of 0.9% whey proteins in distilled water at pH 7.0 at two different heating rates, (——) 0.5°C min⁻¹ and (– – –) ca. 20°C min⁻¹

With knowledge of the gelation mechanisms it may thus be possible
to understand differences in gel structures.

It may, however, be difficult to predict gel structures from
data on denaturation and aggregation. Processing and environmental conditions may influence both chain-chain, molecule-
molecule, and aggregate-aggregate interactions involved in gel
formation and it is not yet known how changes on these levels
affect the final structure.

Many other problems remain to be solved before we can fully
understand structure forming processes. Apart from knowledge
of the gelation mechanisms, where the kinetics involved require
further studies, more work has to be done on the molecular level
as well as on the macroscopic level. The kind and nature of
the crosslinks have to be investigated on the molecular level.
On the macroscopic level it is essential to relate the network
structure with properties like texture and waterbinding. Without such knowledge it will not be possible to gain full insight
into what happens during structure formation and a knowledge
necessary for controlling and optimizing processes important
for food preparation.

Literature cited

(1)   Hermansson, A.-M. (1978). Physico-Chemical Aspects on
      Structure Formation of Soy Proteins. J. Texture Studies
      9, 33-58, 1978.

(2)   Efremov, J.F. (1976). Periodic Colloidal Structures in
      "Surface and Coll. Sci." Ed. Matijevic 8, 85, 1976.

(3)   McKenzie (1971). Milk Proteins II.

(4)   Tanford, C. (1968). Protein Denaturation Part A.
      Adv. Protein Chem., 23, 121, 1968.

(5)   Koshiyama, I. (1972). Comparison of acid-induced confor-
      mational changes between 7S and 11S globulin in soy bean
      seeds. J. Sci. Fd. Agr. 23, 853, (1972).

(6)   Franks, F., and Eagland, D. (1975). The role of solvent
      interactions in protein conformations. CRC Critical Re-
      views in Biochemistry, 3, 165, 1975.

(7)   Catsimpoolas, N., Funk, S.K., and Meyer, E.W. (1970).
      Thermal aggregation of glycinin subunits. Cereal Chem.,
      47, 331, 1970.

(8)   Ferry, J.D. (1948). Protein Gels. Adv. Protein Chemistry,
      4,(1), 1948.

(9)   Tombs, M.P. (1970). Alterations to proteins during pro-
       cessing and the formation of structures in "Proteins as
       Human Food". Ed. Lawrie, R.A., Butterworths.

(10)   Tombs, M.P. (1974). Gelation of Globular Proteins.
       Faraday Discuss. Chem. Soc., 57, 158, 1974.

RECEIVED August 7, 1978.

The Adsorption Behavior of Proteins at an Interface as Related to Their Emulsifying Properties

EVA TORNBERG

Department of Food Technology, University of Lund, Box 740, S-220 07 Lund 7, Sweden

The emulsifying properties of proteins have been a subject of concern for those dealing with functional properties of proteins. The studies so far have been restricted to two main approaches: emulsifying capacity and emulsion stability measurements. The former measures the maximum oil addition until inversion or phase separation of the emulsion occurs, whereas the latter measures the ability of the emulsion to remain unchanged. A variety of empirical methods has been used, which makes it difficult to compare the results obtained by different authors. Not only methods of measurement vary, but also the way the emulsions are prepared, which strongly influences the properties of the emulsions formed (1). As protein stabilized emulsions are usually very stable and the adsorption of proteins at interfaces can be considered as mainly irreversible, the emulsifying properties of the proteins during the emulsification process become increasingly important in determining the properties of the emulsion formed. In this study of the emulsifying properties of three food proteins, this aspect of the emulsification process has therefore been stressed especially since little or no attention has been paid to this variable in earlier studies. The interfacial behavior of the three food proteins has also been studied by measuring the interfacial tension decay and by evaluating the kinetics of adsorption. The interfacial behavior of the proteins was then compared to the behavior obtained from the emulsification studies.

The three proteins chosen for this study are a mildly produced soy protein isolate, kindly provided by Central Soya, a commercially available sodium caseinate (DMV, Holland) and a whey protein concentrate (WPC) obtained by ultrafiltration (UF) and spray drying of cheese whey. Analysis of the proteins is given in (4) and (11). The present protein products have been investigated, when dispersed in distilled water and in 0.2 M NaCl solution at pH 7 denoted as (0 - 7) and (0.2 - 7), respectively.

0-8412-0478-0/79/47-092-105$05.00/0
© 1979 American Chemical Society

Interfacial studies

The interfacial tension decay of the three food proteins at
the air-water interface at 25°C has been monitored with an apparatus based on the drop volume technique. A full description of
the interfacial tension apparatus is given elsewhere (2). The
following procedure was used (for details cf. ref. 3). A drop of
a certain volume, corresponding to a certain interfacial tension
(γ) value, is expelled rapidly, and the time necessary for the
interfacial tension to fall to such a value that the drop becomes
detached is measured. This procedure is repeated for differing
drop sizes, i.e. for different values of the interfacial tension.
A plot of the interfacial tension as a function of time (t) can
then be constructed as seen in Figure 1 for WPC dispersions at
different initial subphase concentrations and ionic strengths.
The rate of reduction of interfacial tension by proteins is determined by three consecutive or concurrent processes: the diffusion of whole protein molecules or aggregates to and attachment
at the interface; spreading or unfolding of already adsorbed molecules on the interface; molecular rearrangements and reconforma-
tions of adsorbed molecules.
The γ-t-curves obtained, as exemplified for WPC in Figure 1,
have been analyzed in terms of distinguishable rate-determining
steps. This analysis can shortly be summarized as follows (cf.
ref. 3). Diffusion is considered to be the rate-determining step
as long as a plot of the interfacial tension against $t^{1/2}$ is linear. When a sufficiently high adsorption energy barrier exists,
the diffusion will no longer be rate-determining, but the rate of
protein penetration into the interface will be rate-limiting.
This barrier consists of the work done ($\Pi\Delta A_{ag}$) in creating an
area ΔA_{ag} in a surface film of surface pressure Π in order to adsorb an active group (ag) in the protein molecule. This situation
is assumed to prevail if a plot of $\ln \frac{d\Pi}{dt}$ versus Π is linear, where
the slope gives ΔA_{ag}. Examples of the two types of plots can be
seen in Figures 2 (4) and 3 (4) for 10^{-2}% (w/w) dispersions of
soy protein, WPC and caseinate at different ionic strengths.
Diffusion controlled adsorption of proteins at an interface
can imply either that the molecules diffuse to the interface and
adsorb without further spreading, or that spreading or unfolding
is so rapid that diffusion becomes the rate-controlling factor.
The penetration step may involve either spreading or unfolding
of already adsorbed molecules, or adsorption of additional molecules arriving at the interface. In order to be adsorbed, however, a surface area ΔA_{ag} has to be cleared in the film, which
requires molecular rearrangements of the already adsorbed protein molecules. The larger the surface area, ΔA_{ag}, to be cleared
the greater the number of residues to be rearranged within the
surface film and the slower is the process.

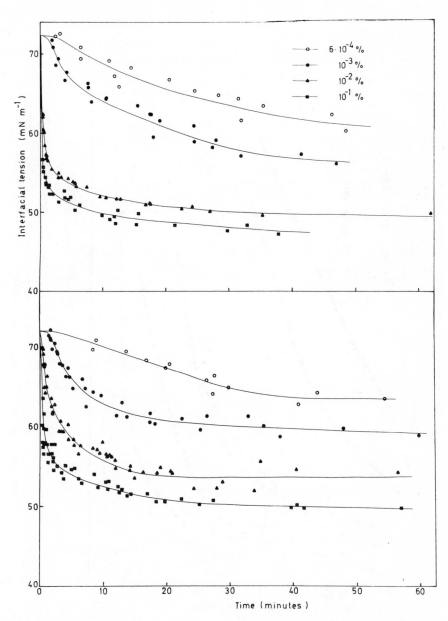

Journal of the Science of Food and Agriculture

Figure 1.   Time dependence of interfacial tensions at the air–water interface for WPC at different initial subphase concentrations. The WPC is dispersed in 0.2M NaCl solution in the upper figure and in distilled water in the lower (4).

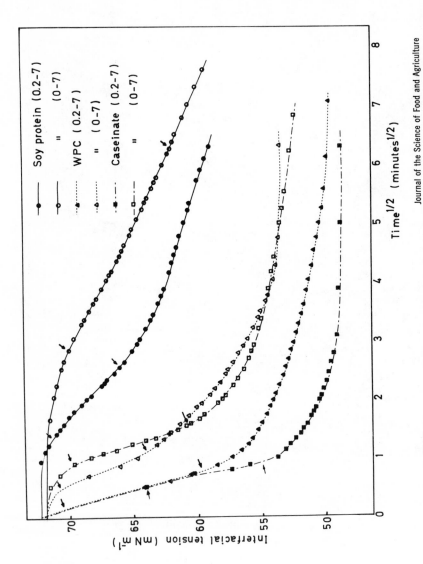

Figure 2. Interfacial tensions at the air–water interface as a function of time$^{1/2}$ for 10^{-2} % (w/w) dispersions of soy protein, WPC, and caseinate at different ionic strengths (4)

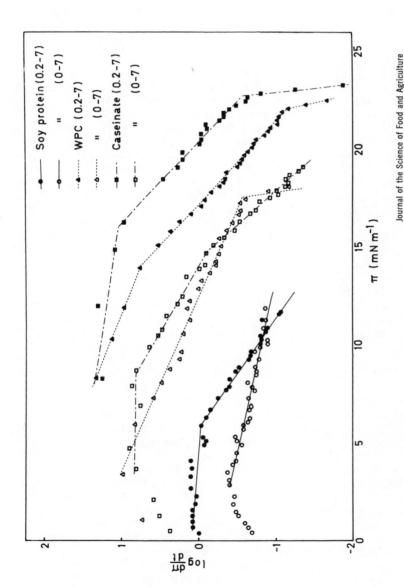

Figure 3.   Log (dΠ/dt) as a function of Π for 10⁻²% (w/w) dispersions of soy protein, WPC, and caseinate at different ionic strengths adsorbing at the air–water interface (4)

Journal of the Science of Food and Agriculture

The concentration dependence of the interfacial tension of the three food proteins can be followed in Figure 4 ($\underline{4}$), where the surface pressure at 40 minutes (Π_{40}) is plotted against the initial subphase concentration. $\Pi_t = \gamma_o - \gamma_t$ (γ_o is the initial interfacial tension). At high concentrations ($10^0 - 10^{-1}$ % (w/w) of initial subphase concentration), the surface activity of all the proteins is high and almost equal, whereas at lower concentrations the differences in surface behavior of the proteins become evident. The caseinate (0.2 - 7) system is most effective as a surface active agent and is more or less independent of concentration in the concentration range of $10^{-1} - 10^{-3}$ % (w/w). The opposite behavior is observed for the soy proteins, which gradually lose their surface activity with decreasing subphase concentration. WPC (0 - 7) and caseinate (0 - 7) have a rather similar concentration dependence in this range, and the curves are in between those of the caseinate (0.2 - 7) and the soy proteins. The addition of 0.2 M NaCl to the WPC dispersions raises the surface activity of the WPC not far beyond that of the caseinate (0.2 - 7). The increase in lowering of interfacial tension due to salt addition is also observed for the other two proteins.

From the kinetic analysis described above of the γ-t-curves of the three food proteins at different ionic strengths and concentrations ($\underline{4}$), the evaluated interfacial behavior is given schematically in Figure 5.

The soy proteins diffuse slowly to the interface in comparison with the other two proteins, which has been interpreted as a result of a higher particle weight of the migrating unit in the case of the soy proteins. This is in accordance with the fact that the soy proteins, consisting of mainly of the 7S and 11S globulins, have a complex quaternary structure in bulk with a particle weight ranging from 180 - 363 000 ($\underline{5,6}$). According to Wolf ($\underline{7}$) and Koshiyama ($\underline{8}$) association and aggregation of the 7S and 11S globulins are favoured at 0.1 - 0.2 M NaCl, which has been illustrated in Figure 5. Contrary to what could be expected, the diffusion of soy protein (0 - 7) is slower than for soy protein (0.2 - 7), which might be explained by a reduced electrostatic repulsion during adsorption for soy protein (0.2 - 7) and thereby a higher net collision rate. The rate-determining step of penetration of protein molecules or segments into the surface film starts at a relatively low surface pressure for the soy proteins, especially in 0.2 M NaCl, compared to the other proteins. This indicates a high spreading ability of the already adsorbed soy protein aggregates, which is most pronounced when the soy protein is dispersed in 0.2 M NaCl solution.

Although the caseinate has a complex quaternary structure, like the soy proteins, it has a very different surface behavior. The diffusion step is almost as rapid as that of WPC at concentrations above 10^{-3} % (w/w). This can be explained if the migration to the interface of the caseinates is performed mainly by the

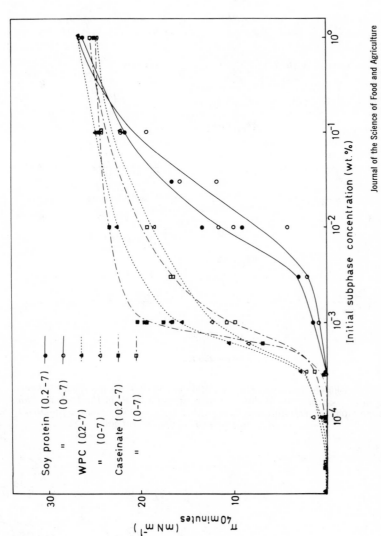

Journal of the Science of Food and Agriculture

Figure 4. The surface pressure attained after 40 minutes ($\Pi_{40\,minutes}$) as a function of the initial subphase concentration for all the proteins studied at different ionic strengths (4)

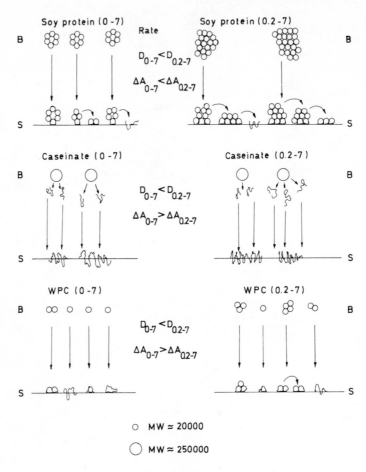

Figure 5. A highly schematic representation of the differing interfacial behavior of the three food proteins at different ionic strengths. (B) in the bulk medium; (S) at the interface; (D) diffusion rate; (ΔA) is assumed to be inversely related to the rate of rearrangements in the protein film.

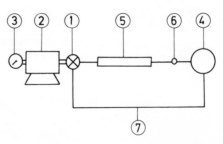

Figure 6. A representation of the isothermal recirculating, emulsification system (10). (1) Gear pump; (2) reversible motor; (3) revolution counter; (4) emulsifying chamber; (5) heat exchanger; (6) thermometer; (7) plastic tubings.

free casein molecules, which are in equilibrium with those in
the submicelles. The results obtained from the work by Creamer
and Berry (9) show that the casein 'micelle' subunits with a molecular weight of approximately 250 000, which sodium caseinate
is supposed to consist of to a large extent, are in equilibrium
with their components caseins. The diffusion-controlled occupation
of the interface is very evident for the caseinates, especially
at an ionic strength of 0.2, compared to the other proteins. The
present results indicate that the very flexible, random coil-like
casein molecules have simpler kinetics than the other proteins
studied, involving diffusion to the interface and direct ancho-
ring of the freely available hydrophobic segments. Due to higher
electrostatic repulsion between adsorbed species at the interface
for caseinate (0 - 7) compared to caseinate (0.2 - 7), the caseinate (0.2 - 7) molecules probably can pack at the interface
with greater density. This will give rise to an adsorption barrier
at a higher Π for caseinate (0.2 - 7), which has also been observed. The reduced speed of the diffusion process of the caseinate
(0 - 7) as compared to caseinate (0.2 - 7) might be explained by
an enhanced electrostatic repulsion during the adsorption of
caseinate (0 - 7) and thereby a reduction of net collision rate.

   The whey proteins diffuse quickly to the interface, which
is in accordance with the fact that the whey proteins consist
mainly of small molecules and molecular complexes. The WPC
(0 - 7) diffuses somewhat slower than WPC (0.2 - 7) to the interface, the same relationship which has been observed for the other
two proteins. WPC (0.2 - 7) covers the interface to a higher degree by diffusion than WPC (0 - 7), which indicates that WPC
(0 - 7) spreads or unfolds more easily at the interface.

Emulsification studies

   In order to evaluate the emulsifying properties of proteins
as a function of the emulsification variables, it is important
to quantify and to describe more exactly the emulsification process. This has been done by utilizing a recirculating, emulsification system (10), where the flow velocity is controlled. The
system is schematically represented in Figure 6, and consists of
a gear pump 1 connected to a reversible motor 2 and a revolution counter 3 , an emulsifying chamber with an inlet and an out-
let 4 , a heat exchanger 5 with a thermometer and plastic tubings, which complete the recirculating system. The emulsifying
part 4 can be varied from emulsification with a turbomixer to
sonication and to valve homogenization. The power and energy input during the emulsification process have been measured (10),
thereby providing a tool for the comparison of the emulsifying
efficiency of various kinds of emulsifying equipment.

   The protein stabilized emulsions formed were made up of
40 % (w/w) soybean oil and 60 % (w/w) protein dispersion of 2.5
% (w/w) protein content. A quantity of 50 grams was emulsified

in all the experiments, and 30 grams of the emulsion formed were
stored for 24 hr at 20°C after formation. The emulsions were
characterized in terms of fat particle size distribution and the
amount of protein adsorbed per unit area of fat surface (protein
load). The methods used are fully described elsewhere (11). For
the particle size analysis a spectroturbidimetric method has been
used, in which the average globule size obtained is the arithme-
tic mean of the surface-weighted distribution, denoted as d_{vs}.
It relates directly to the specific surface area, A, by the
following formula:

$$A = \frac{6\ \Phi}{d_{vs}}\quad m^2\ fat/ml\ emulsion$$

where Φ is the volume fraction of the dispersed phase. So far only emulsions made with a valve homogenizer incorporated into the
recirculating system have been characterized with regard to the
particle size distribution and protein load.

In Figure 7 the fat surface area, A, obtained after 10 passes
in the recirculating system is plotted against power input during
valve homogenization for all the protein stabilized emulsions
(11). The surface area of the more or less non-flocculated emulsions, such as the caseinate (0.2 - 7), WPC (0 - 7) and caseinate
(0 - 7) stabilized emulsions, show a similar and almost linear
increase with power input. The degree of flocculation was estimated by microscopic evaluation. At lower power consumptions the
soy protein (0 - 7), WPC (0.2 - 7) and soy protein (0.2 - 7) stabilized emulsions, which are flocculated after homogenization,
produce larger fat surface areas than the non-flocculated ones.
This is most pronounced for the soy protein (0.2 - 7) stabilized
emulsions, which are heavily flocculated and viscous compared to
the other protein stabilized emulsions. The relative width of the
fat droplet size distribution, c_s, is plotted against power input
during valve homogenization at 10 passes in Figure 8 (11). It is
seen in the figure that the flocculated emulsions give emulsions
with larger spread in globule size than the non-flocculated systems. No discernible difference within the limits of error in the
c_s-power-dependence can be observed among the non-flocculated
emulsions stabilized by different proteins.

Figures 9 and 10 show the fat surface area and the relative
width of the globule size distribution, respectively, of all the
protein stabilized emulsions as a function of the number of passes
at a power consumption of 40 W (11). As can be seen from Figures
9 and 10 an increase in number of passes does not noticably enhance the final fat surface area of the protein stabilized emulsions, but affects more the distribution width by a decrease and
a final level off. The caseinates at both ionic strengths produce
emulsions of the smallest surface area and the narrowest distribution widths. The WPC stabilized emulsions have larger surface
areas and wider spread in globule size than emulsions stabilized

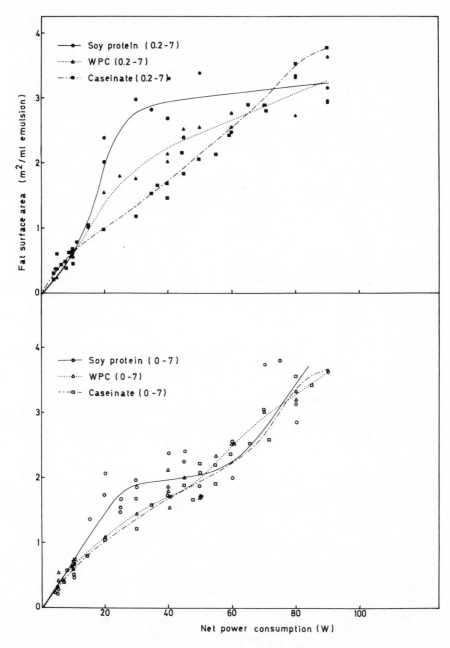

Figure 7.   Fat surface area of the protein-stabilized emulsions as a function of power consumption after emulsification with a valve homogenizer using 10 passes through the recirculating system (11)

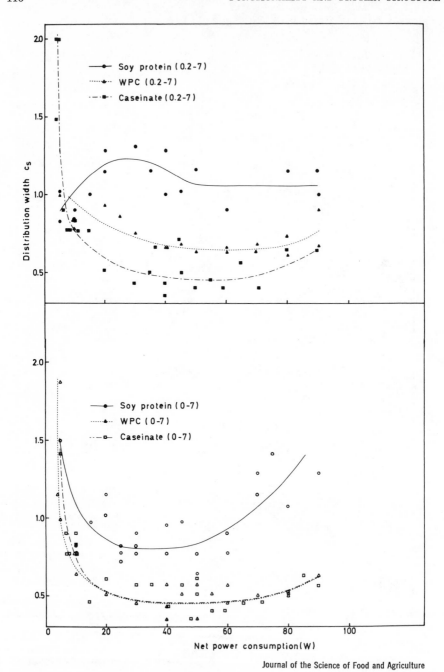

Figure 8.   Distribution width (c$_s$) of the protein-stabilized emulsions as a function of power consumption, on emulsification with a valve homogenizer using 10 passes through a recirculating system (11)

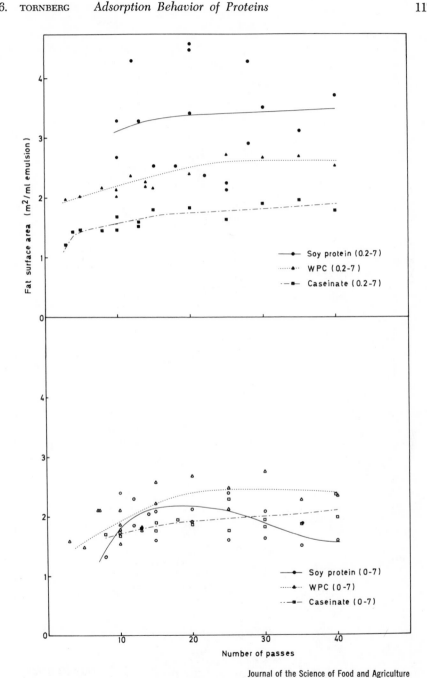

Figure 9. Fat surface area of the protein-stabilized emulsions as a function of number of passes after emulsification with a valve homogenizer at constant power supply of 40 W (11)

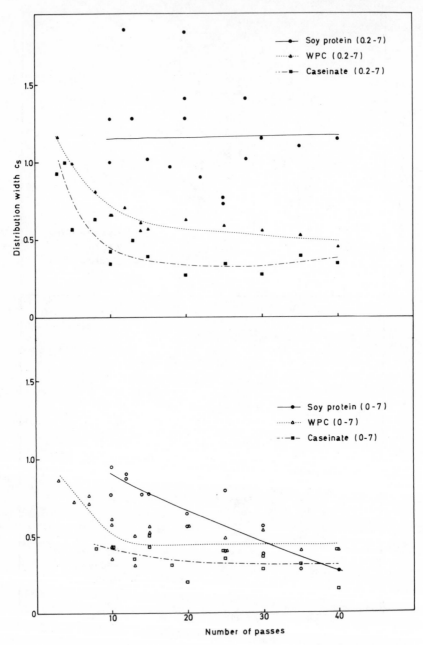

Figure 10.   Distribution width (c_s) *of the protein-stabilized emulsions as a function of number of passes, on emulsification with a valve homogenizer at a constant power supply of 40 W (11)*

by caseinate, where the more flocculated WPC (0.2 - 7) stabilized emulsions have larger distribution widths. The soy protein (0 - 7) stabilized emulsions have a rather similar dependence on number of passes as for the caseinate and the WPC stabilized emulsions, whereas no direct relationship can be drawn from the results obtained for the soy protein (0.2 - 7) stabilized emulsions, due to the wide scatter in results.

In Figure 11 the amount of protein adsorbed in mg per m^2 fat surface area (protein load) is plotted against the fat surface area created by augmenting the power input in accordance with Figure 7. The most striking feature to be observed is the very different behavior shown by the caseinates as compared to the soy proteins. The emulsions stabilized by the latter have a very high protein load at small surface areas, which diminishes almost exponentially as the surface area expands. The caseinates, though, have a low value of protein load for emulsions of small surface area, but it grows as a function of the surface area to a maximum at an area of about 1.5 m^2/ml emulsion, after which it decreases slightly. For both these proteins the addition of salt results in more protein adsorbed at the interface for all the surface areas investigated.

The WPC (0.2 - 7) stabilized emulsions have the lowest protein load ($\approx$ 1.5 mg/m^2) at fat surface areas between 1.0 and 3.0 m^2/ml, whereas at larger surface areas, the soy protein (0 - 7) stabilized emulsions have as low values as those stabilized with WPC (0.2 - 7). It is interesting to note that an increase in ionic strength to 0.2 M NaCl does not increase the amount of protein adsorbed in the case of the whey proteins. In fact the opposite is observed, in contrast to the behavior of the other two proteins.

In Figure 12 the percentage of adsorbed protein from bulk is plotted as a function of the fat surface area created by increasing the power input during valve homogenization. The very different protein load obtained by the caseinates and the soy proteins in Figure 11 corresponds in Figure 12 to a higher percentage adsorbed by the soy proteins at small surface areas and a lower percentage adsorbed at the larger surface areas as compared to the caseinates. It can also be deduced from Figure 12 that, in order to cover the interface with protein at larger surface areas, the caseinates, especially in 0.2 M NaCl, predominantly supply protein from the bulk. The soy proteins, though, do so to a lesser extent, especially at larger areas, where percentage protein adsorbed become more independent of fat surface area. This indicates that, at these large surface areas, the newly created interface is mostly covered by spreading or unfolding of already adsorbed soy protein molecules at the interface. This emulsifying behavior of the two proteins seems to be in accordance with the interfacial behavior suggested from the interfacial tension measurements. The high protein coverage of the soy proteins at small surface areas is in accordance with the sugges-

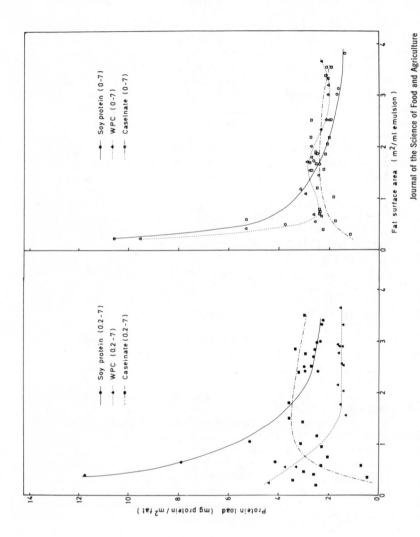

Journal of the Science of Food and Agriculture

Figure 11. Protein load of the protein-stabilized emulsions as a function of the fat surface area created during variation of power supply at a constant number of passes of 10 (11)

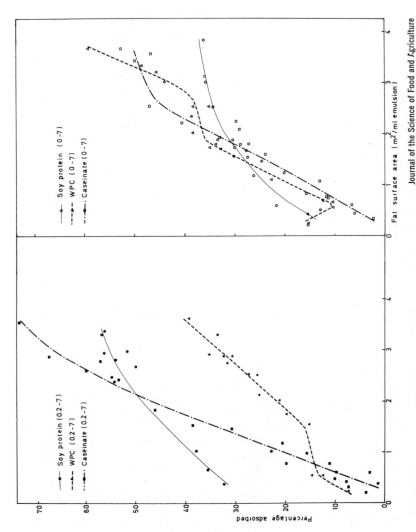

Journal of the Science of Food and Agriculture

Figure 12. Percentage adsorbed protein from the bulk to the fat interface of the protein-stabilized emulsions as a function of the fat surface area created during variation of power supply in a valve homogenizer using a constant number of passes of 10 (11)

tion that the soy proteins migrate to the interface by the gross associated complex, whereas the low percentage adsorbed for the caseinates at these surface areas support the assumption that casein monomers are adsorbed at the interface. Moreover, the interfacial tension results indicated that the soy proteins spread relatively easy at the interface, especially in 0.2 M NaCl solution, and this is mainly the way the soy proteins cover the enlarged interface during emulsification, as suggested from Figure 12. The percentage caseinate adsorbed is more or less directly proportional to the fat surface area as seen in Figure 12, which suggests that, the larger the fat surface area created, the more casein monomers are extracted from the casein subunits. This behavior is especially pronounced, when caseinate is dispersed in 0.2 M NaCl solution.

The emulsifying behavior of the whey proteins is not as easily correlated with the interfacial behavior as that of the other proteins. The interfacial data suggested that WPC (0 - 7) spreads relatively easy at the interface in relation to WPC (0.2 - 7), but in Figure 12 the supply of molecules from the bulk by WPC (0 - 7) exceeds that of WPC (0.2 - 7). This might be explained if the more frequently occurring unfolding of the WPC (0 - 7) molecules gives rise to an enhanced amount of hydrophobic groups exposed to the bulk, which induces further adsorption of protein from the bulk and, hence, multilayer formation.

Summary

The interfacial and emulsifying behavior of three food proteins, a soy protein isolate, a sodium caseinate and a whey protein concentrate (WPC) have been studied. A kinetic analysis of the interfacial tension decay of the proteins indicates the following characteristics. The soy proteins diffuse slowly to the interface compared to the other proteins, probably with the quaternary structure intact, which disintegrates when adsorbed at the interface. Both the whey proteins and the caseinates diffuse quickly to the interface, where for the caseinates the diffusion--controlled occupation of the interface is very evident, especially at an ionic strength of 0.2.

When preparing protein stabilized emulsions in a valve homogenizer, the final fat surface area is mostly determined by the conditions during emulsification, and if the emulsion system flocculates or not. The final fat surface area of the emulsions obtained increases more as a function of power input than as a function of number of passes. Distribution width, c_s, decreases mostly with increasing power supply and number of passes, but at the highest power input c_s often starts to increase again. Flocculation of emulsions gives generally larger spread in globule size. The amount of protein adsorbed on the fat globule (protein load) is largely determined by the fat surface area created and by the type of protein adsorbed. Salt addition to 0.2 M NaCl enhances protein adsorption at the fat globule interface in the case of soy protein and caseinate, but for the whey proteins protein load is higher in distilled water. The interfacial behavior of the soy proteins and the caseinates suggested from the interfacial tension measurements seem to be in accordance with the emulsifying behavior, whereas the emulsifying behavior of the whey proteins is not as easily correlated with the interfacial behavior.

Literature cited

1. Tornberg, E. and Hermansson, A.-M. J. Food Sci., (1977) 42 (2) 468.
2. Tornberg, E. J. Coll. Interface Sci., (1977) 60 (1) 50.
3. Tornberg, E. J. Coll. Interface Sci., (1978) 64 (3) 391.
4. Tornberg, E. Accepted for publication in J. Sci. Fd Agr.
5. Koshiyama, I. Agr. Biol. Chem., (1971) 35, 385.
6. Badley, R.A., Atkinson, D., Hauser, H., Oldani, P., Green, J.P., and Stubbs, J.M. Biochim. Biophys. Acta (1975) 412, 214.
7. Wolf, W.J. In "Soybeans: Chemistry and Technology", (Smith, A.K., and Circle, S.J., eds.), 1972, p. 93, AVI Publishing Co. Inc., Westport, Connecticut.
8. Koshiyama, I. Agr. Biol. Chem., (1968) 32, 879.
9. Creamer, L.K., and Berry, G.D., J. Dairy Res., (1975) 42, 169.
10. Tornberg, E. and Lundh, G. J. Food Sci (1978) 43, 1553.
11. Tornberg, E. Accepted for publication in J. Sci. Fd Agr.

RECEIVED October 23, 1978.

The Influence of Peptide Chain Length on Taste and Functional Properties of Enzymatically Modified Soy Protein

J. ADLER-NISSEN and H. SEJR OLSEN

NOVO INDUSTRI A/S, Enzyme Applications Research and Development, Enzymes Division, Novo Allé, DK-2880 Bagsvaerd, Denmark

Enzymatic hydrolysis of food proteins generally results in profound changes in the functional properties of the proteins treated. Protein hydrolysates may therefore be expected to fulfil certain of the food industry's demands for proteins with particular, well-defined functional properties. A wide-spread use of protein hydrolysates in food requires, however, a careful control of the taste and functionality of the protein during its hydrolysis and subsequent processing to obtain a reproducible product quality.

The composition of a protein hydrolysate is conveniently described by the degree of hydrolysis (DH), which is defined as the percentage of peptide bonds cleaved (1). The average peptide chain length (PCL), measured in number of amino acid residues, can be shown to be related to DH by the following equation:

$$DH = (\frac{1}{PCL} - \frac{1}{PCL_O}) \times (1 + \frac{1}{(PCL_O-1)}) \times 100\% \qquad (1)$$

where PCL_O is the average peptide chain length of the intact protein. Both PCL and PCL_O are the number-average peptide chain lengths.

In most cases PCL_O is large, $PCL_O \gg PCL$ and (1) approaches

$$DH \simeq \frac{100\%}{PCL} \qquad (2)$$

Many workers have studied the influence of enzymatic hydrolysis on the functional properties of various food proteins, and much of this work has recently been reviewed by Richardson (2). However, there seem to be very few reports which quantitatively relate functionality to parameters which characterize the protein hydrolysates per se (e.g. molecular weight). Ricks et al. (3) examined the solubility and taste of a number of pure proteins (denatured pepsin, lactoblobulin, α-S_1-, κ-, and β-casein) hydrolysed with

0-8412-0478-0/79/47-092-125$05.50/0

various proteolytic agents. They found a weak, but significant
correlation (r = 0.6) between the solubility at pH 4.5 and DH.
They also observed that the bitter taste of the peptides depended
on many parameters; DH being one of them. It seemed clear from
their work that non-bitter hydrolysates could be obtained at high
DH-values (above 20%); an observation which is in accordance with
the results of Clegg and McMillan (4) who removed the bitter
taste of casein hydrolysates by applying an exopeptidase.

The present work was carried out with the purpose of relating
taste, solubility, emulsifying capacity, foaming capacity, and
viscosity of soy protein hydrolysates to the DH of these hydrolysates. This tentative approach to the manufacture of functional
soy protein hydrolysates was chosen, because DH is easily controlled during hydrolysis by means of the pH-stat technique (5),
and because the properties of the hydrolysate are presumed to be
related to the DH-value rather than to hydrolysis parameters
such as temperature, substrate concentration, and enzyme-substrate ratio (6).

Materials and Methods

Materials. The soy protein isolate used in all experiments
was Purina 500 E from Ralston Purina, a type of nonhydrolysed soy
isolate which is generally used in the European meat industry.
Gelatin (alkaline extracted) was obtained from Extraco, Sweden.
The enzymes used were Alcalase[R] 0.6 L, a liquid, food-grade
preparation of subtilisin Carlsberg, and Neutrase[R] 0.5 L, a
liquid, food-grade preparation of a B. subtilis neutral protease.
Both enzymes are commercially available from Novo Industri A/S
(7). All other reagents were analytical grade laboratory chemicals.

Hydrolysis Curves. The hydrolysis curves, i.e. graphs describ-
ing the DH/time relationship, for soy protein isolate hydrolysed
with Alcalase and Neutrase, respectively, were obtained by the
pH-stat technique, using the following hydrolysis parameters:
Substrate conc. (S) = 10% protein (N x 6.25); enzyme-substrate
ratio (E/S) = 2.0%; pH 8.0 (Alcalase) and pH 7.0 (Neutrase), respectively; temperature (T) = 50°C; total mass of reaction mixture (M) = 800 g. The hydrolyses were carried out for at least
5 hours in a 1000 ml reaction vessel, equipped with a stirrer, a
thermometer, a pH-electrode, and a plastic tube for the admission
of base. The pH-stat arrangement consisted of standard equipment
from Radiometer, Copenhagen, (pH-meter, titrator, autoburette,
and recorder). The hydrolysis curves were calculated from the
base consumption by means of the following equations:

$$h = (1 + 10^{(pK-pH)}) \times \frac{N_b \times B}{M \times (S\%/100)} \tag{3}$$

and

$$DH = \frac{h}{h_{tot}} \times 100\% \tag{4}$$

where h = "hydrolysis equivalent", defined as meqv. peptide
bonds cleaved per gram protein (N x 6.25)

B = base consumption in ml

pK = average pK-value of the α-amino group in the peptides

N_b = normality of the base

h_{tot} = total number of peptide bonds in a protein, given in
meqv. per gram protein (N x 6.25).

Equation (3) can be deduced from the theoretical discussion of
the pH-stat by Jacobsen et al. (8) and equation (4) follows immediately from the definition of DH (1). We have measured a tentative pK-value of 7.1 at 50°C (9); this value is in accordance
with the theoretical value which can be calculated from the data
of Steinhardt and Beychock (10). h_{tot} is calculated from the amino acid composition of the soy isolate by summing up the contents
(in mmol per gram protein) of the individual amino acids. For
this particular isolate, h_{tot} = 7.75 meqv./g (9).

Production of Hydrolysates. The hydrolysates of soy protein
isolate were produced using exactly the same hydrolysis conditions as used above for obtaining the hydrolysis curves. Nine
hydrolysates covering a range of DH-values were produced (four
with each of the two enzymes and one with no enzyme added to
serve as a control). In all cases the hydrolysis was terminated
by the addition of HCl to pH 4.2 to inactivate the enzyme (5).
pH was then adjusted to pH 7.0 using NaOH. NaCl was added until
the final concentration in a 10% protein (N x 6.25) solution was
0.25 M NaCl. The hydrolysates were then freeze-dried, analysed
for nitrogen content (semi-micro Kjeldahl), and content of free
amino groups to determine the exact DH-value of the individual hydrolysates. The free amino groups were assayed by reaction with
trinitrobenzene sulphonic acid (TNBS) using L-leucine as internal standard, and DH was determined from a linear standard curve
which relates L-leucine equivalents per gram protein (N x 6.25)
to DH. This method of DH determination is described in detail
elsewhere (11).

In addition to the 9 hydrolysates mentioned above, 4 soy protein hydrolysates (plus 1 control) were produced on a pilot plant
scale, using Alcalase (5, 6). All processing conditions were
identical to those mentioned previously except that the enzyme
inactivation was performed by pasteurisation at 90°C for 30 sec.
in an Alfa-Laval small-scale plate pasteuriser P-20 with a flow
of 2.6 l/min. Ten litres of hydrolysate were pasteurised, freeze-
dried, and the DH determined as described above.

The gelatin hydrolysates were produced in 5000 ml vessels
using the following hydrolysis parameters: S = 30% dry matter
(corresponding to 26.6% protein (N x 5.55)); pH 5.0 - 5.1; T =
55°C; mass of reaction mixture = 4000 g. The enzymes used were
Alcalase, Neutrase, and an 1:1 mixture of both in varying concentrations (E/S = 1 - 4%). During hydrolysis, 100 ml samples were
drawn and rapidly heated to 90°C to inactivate the enzyme. DH of
each sample was determined (11) and the viscosity was measured in

a rotation viscosimeter as described below.

Taste Evaluation. The four hydrolysates produced in pilot plant were evaluated for bitter taste by the laboratory's taste panel. Tasting took place in the taste panel room which is equipped with separate booths, and the panel has been selected and trained specially for discrimination of bitterness. The panel was instructed to rank two samples and four bitter-tasting standards, containing 20, 40, 80, and 160 ppm quinine hydrochloride dissolved in non-bitter iso-electric soluble soy protein hydrolysate (5, 6). 20 ppm quinine hydrochloride in this solvent had in previous experiments been established as the panel's threshold value. The protein (N x 6.25) concentration in the samples and standards was 4.0% and pH was adjusted to 6.5 with 4 N NaOH or 6 N HCl. Thus, the test consisted of six individual items, which had to be ranked according to increasing bitterness. The sum of ranks (SR) for each of the six items was calculated and used to assign a quinine equivalent value (QEV) to the samples by the following method: The SR-values for the four standards were plotted against the logarithm of the quinine concentration (in ppm) and the four points were connected with straight lines. The abscissae of the inter-sections between the SR-values for the samples and these connecting lines were then used as QEV-values (in ppm). SR-values outside the range of standards cannot be assigned a QEV.

The test was carried out twice so that all four samples could be evaluated. A week later the four samples were evaluated once more by the described procedure.

The individual panelists' ranking of the four standards should not be influenced by the presence of two extra, clearly identifiable items, and the panel could therefore be tested for its ability to rank the standards correctly. This was done by calculating the "coefficient of concordance", as described by e.g. Moroney (12). The SR-values used in these calculations were the SR-values obtained after omitting the two samples from each individual ranking and the null-hypothesis for the test was that the actual SR-values should be equal to the SR-values which would have been obtained if all the panelists had ranked the standards correctly.

Because of the inherent salt content of the hydrolysates produced in the laboratory, these were not evaluated organoleptically.

Gel Chromatography. Gel chromatography was performed on Sephadex G 50 in a Pharmacia 26 x 1000 mm thermostated column with a bed volume of approximately 500 ml. The eluent flow was controlled by a peristaltic pump (Pharmacia P-3) and the effluent was monitored at 206 nm by a UV detector (LKB Uvicord III) connected to a galvanometric recorder (LKB 6520). The eluent was a phosphate buffer (0.0325 M K_2HPO_4, 0.0026 M KH_2PO_4, 0.40 M NaCl) of pH 7.6 and ionic strength 0.5. This buffer was used by Arai et al. (13) in gel chromatography of soy protein hydrolysates. The sample was dissolved in the buffer and filtered before application to the column.

Nitrogen Solubility Curves. Nitrogen solubility of the soy

samples was determined over a pH range of 2.0 - 9.0 in a 1% protein dispersion in 0.2 M NaCl by the following procedure:

2.0 g of protein product was dispersed in 150 ml of 0.2 M NaCl
using a laboratory blender for two minutes. The blender was washed
with 50 ml of 0.2 M NaCl and the washing liquid combined with the
homogenised sample. pH was adjusted with 0.2 M HCl or 0.2 M NaOH
and the dispersion was stirred with a magnetic stirrer for 45 minutes. pH was regularly adjusted, if necessary. At the end of the
stirring period, the volume was determined by weighing and 25 ml
of the dispersion was centrifuged at 4000 x g for 30 minutes. The
supernatants were analysed for nitrogen content by the Kjeldahl
procedure (double determination), and the nigrogen solubility was
calculated as (soluble N%/total N%).

Emulsifying Capacity. The method of Swift et al. (14) was
used with slight modifications to determine the emulsifying capacity of the soy samples:

Two grams of protein (N x 6.25) were dispersed at low speed
for two minutes using 200 ml of 0.5 M NaCl. 25 ml was transferred
to a blender jar and weighed. 25 ml of soy bean oil was added and
the oil-water mixture was subjected to high-speed cutting and
mixing (approximately 13,000 rpm)with a MSE Homogenizer. The
blender jar was cooled in an ice bath. During this mixing a steady
flow of oil was added from a separating funnel at a rate of 0.3 ml
per sec. (The rate of oil addition has an effect on the emulsifying capacity). Oil was added until the emulsion formed resisted
mixing. Then the position of the blades of the homogenizer was
changed in order to secure the absorption of fresh oil. This
thickening indicates that the "end point" is about to be reached.
Addition of oil was immediately terminated when a sudden decrease
in viscosity was observed visually. This indicates that the emulsion had collapsed from an oil/water emulsion to a water/oil emulsion. The total amount of oil added before the "end point" was
found by weighing the blender jar. Emulsifying capacity was calculated as ml oil per gram protein (N x 6.25). The density of the
oil was taken as 0.9 g/ml. The emulsifying capacity was measured
five times for each product. The average coefficient of variation
was 6.96% on the final results based on 5 x 10 determinations.

Whipping Expansion. The method of Eldridge et al. (15) was
used with slight modifications to determine the percentage of expansion:

500 ml of a 3% (N x 6.25) aqueous dispersion of the protein
samples were whipped at speed III for 4 minutes in a Hobart mixer
(model N-50) mounted with a wire whip. The whipping expansion was
calculated as follows:

$$\text{Expansion} = 100\% \text{ x } \frac{V-500}{500}$$

where V = final whip volume in ml.

V was measured by refilling the mixer jar with water. Because
of the fairly large consumption of protein hydrolysate used in

this procedure, the whipping expansion could only be measured once on each product.

Viscosity. Soy protein hydrolysate dispersion containing 10% protein (N x 6.25) and 0.5 M NaCl were prepared using a blender during repeated periods of 30 sec. until the slurry was homogenous. The slurry was centrifuged at 500 rpm for 3 min. in order to remove dispersed air and finally gently stirred with a spatula.

Heat treatment was performed in closed containers in a water bath maintaining a sample temperature of 80°C for 30 sec.

The viscosity of the gelatin hydrolysates was measured directly on the samples.

Flow curves were drawn at 25°C (30°C for the gelatin hydrolysates) using a HAAKE-roto-viscosimeter RV2 with measuring systems MVI and MVII. By means of a PG 128 programmer connected to the viscosimeter the speed was increased from 0 to 512 min.$^{-1}$ for 40 sec. and reversed at the same rate. This programme was repeated three times with 5 sec. holding time in between. The viscosity was determined using the slope of the flow curves between 256 and 512 min.$^{-1}$. This seemed to be a reasonable procedure because all the hydrolysates exhibited approximately Newtonian behaviour, judging from the virtually linear flow curves.

Results and Discussion

Hydrolysis Curves. Figure 1 shows the two hydrolysis curves obtained by hydrolysis of Purina 500 E with Alcalase and Neutrase, respectively. The difference in kinetics between the two enzymes is apparent from the difference in shape of the two curves. The hydrolysis curves also serve to indicate the hydrolysis time needed to reach a desired DH-value.

Bitterness. Table I shows the results from the ranking of the four hydrolysates according to increasing bitter taste. It is obvious that bitterness and DH are positively correlated - a result which, at first, appears to be in contradiction to the results of Ricks et al. (3). However, the DH-values shown here are small compared with those of Ricks et al. - and besides, the taste of the intact protein is non-bitter, so the bitterness must rise at the beginning of the hydrolysis. The relationship between bitterness and DH may therefore be expected to look like Figure 2; further experiments may confirm this.

To explain the shape of Figure 2, the following assumptions are made: 1) Bitterness is related to the hydrophobicity of the peptides, as first proposed by Ney (16). The higher the content of hydrophobic amino acids in the peptide chain, the greater the tendency to get bitter tasting peptides. Ney's hypothesis seems to be generally acknowledged (3, 17, 18). 2) To exert influence on the bitterness, the hydrophobic side-chains must, as stated by Matoba and Hata (19), be exposed to the solvent so that reaction with the receptors in the taste buds can take place. Support for this assumption can be found in the well-known debittering effect of the plastein formation (20). Although the mechanism for plastein reaction is not fully known, it appears that hydrophobic

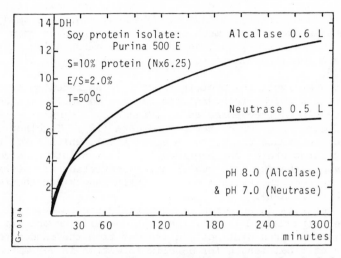

Figure 1.   Hydrolysis curves for soy protein isolate hydrolyzed with alcalase and neutrase, respectively

Table I. Results from the Organoleptic Evaluation of the Bitterness of Four Soy Protein Hydrolysates

Soy protein hydrolysate[1] DH		3.6	5.3	7.1	8.7
Quinine equivalent value, QEV ppm	1. week	<20	24	48	78
	2. week	<20	<20	48	140
Coefficient of concordance for standards, W	1. week[2]	0.148	0.028	0.148	0.028
	2. week[3]	0.054	0.005	0.054	0.005

[1]) The four hydrolysates were produced in pilot-plant as described in the experimental section, and an alternative procedure (5,6) would have yielded much lower levels of bitterness. The hydrolysates were evaluated at pH 6.5 and in a concentration of 4% protein (Nx6.25).
[2]) 10 panelists. 90% significance level for W: $W \gtreqless 0.211$, tested according to Moroney (12). The low values of W indicate in this case a good agreement between the judges. W = 0 denotes complete agreement with the correct ranking of the standards.
[3]) 9 panelists. 90% significance level for W: $W \gtreqless 0.234$, testing and comment as above.   G~0420

interactions take place in plasteins (21), thus masking the side-chains responsible for bitterness. 3) Hydrophobic amino acids, whose amino and carboxylic ends are blocked, e.g. by forming peptide linkages, contribute much more to the bitterness than if one of these ends are free, such as in terminal amino acids (19). These again contribute more to the bitterness than do free amino acids (19).

During hydrolysis the protein is rapidly degraded to intermediate peptides which are hydrolysed further to small peptides and sometimes even to free amino acids (Figure 3; a simpler version of this figure is given by Matoba and Hata (19)). Together with the three statements above, the present model, which takes the secondary structure of the peptides in solution into consideration, gives a theoretical, qualitative description of the relationship between the average peptide chain length and the formation of bitter taste.

Small peptides in solution are generally random coils; however, above a certain critical length, the peptides will be able to have secondary structures distinctly different from the random coil. Thus, the critical length for α-helix formation is 7-9 amino acid residues (22). It is not generally possible to predict at what chain length two hydrophobic side-chains are able to interact, whereby the peptide becomes U-shaped, because this must depend on the actual position and nature of the side-chains. This hydrophobic interaction masks the side-chains, resulting in a reduction of the bitter taste.

The description above bears some resemblance to the reversible folding/unfolding of proteins. A speculative conclusion from this analogy would be that an increase in temperature should be followed by an increased exposure of hydrophobic side-chains. Hypothetically, an increase in bitter taste, relative to a standard quinine solution, should be observed at elevated temperatures; whether this is found in practice or not remains to be investigated.

To summarize the model above: At low DH-values, the majority of the hydrophobic side-chains are still masked and bitterness is low. At increasing DH-values more and more peptides will be too small to form proper secondary structures, the hydrophobic side-chains become exposed, and bitterness increases. At still higher DH-values, however, the peptides are so small that a significant fraction of hydrophobic amino acid will be either free or in terminal position and this will tend to reduce the bitter taste.

Solubility. The results from the solubility experiments are given in Figures 4 and 5. The broken curve is the solubility curve for the original soy protein isolate (Purina 500 E); the nitrogen solubility of this product at neutral pH is below 40%, which indicates that the isolate has been partly denatured during its processing. The sample denoted as DH = 1.0% is the control and it is clearly demonstrated that the acid treatment has caused further aggregation and denaturation. The definitely positive DH-value

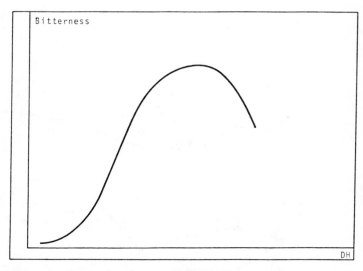

Figure 2.   *Expected qualitative relationship between bitterness and DH*

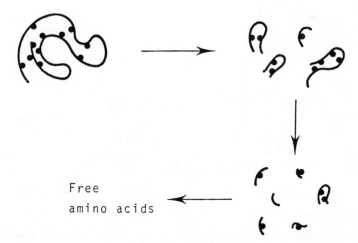

Figure 3.   *Exposure of hydrophobic regions during enzymic degradation. The black circles denote hydrophobic side chains.*

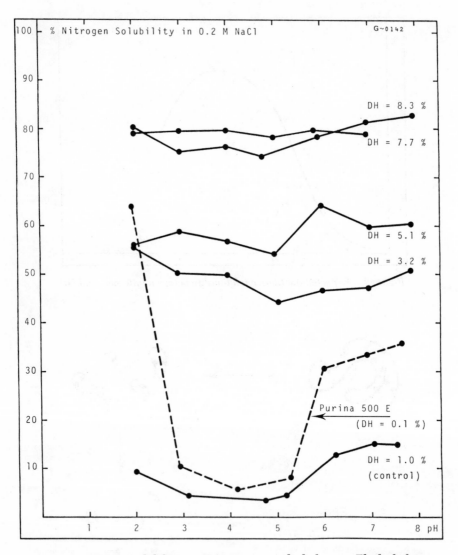

Figure 4.   Nitrogen solubility vs. pH for soy protein hydrolysates. The hydrolyses were carried out using alcalase[(R)] at pH 8.0 as described in the experimental section.

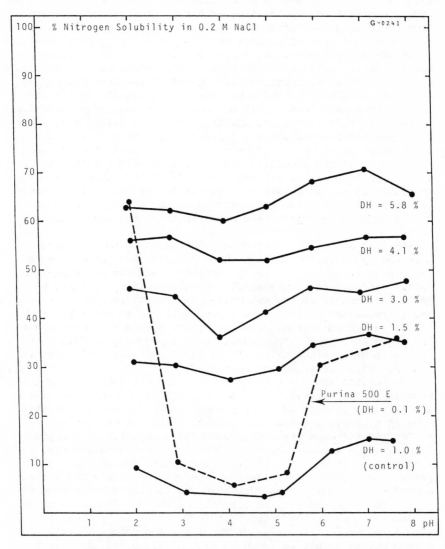

Figure 5.   Nitrogen solubility vs. pH for soy protein hydrolysates. The hydrolyses were carried out using neutrase[R] *at pH 7.0 as described in the experimental section.*

may be due to an increase in the number of available free amino
groups, as a result of the unfolding of the protein.

For both enzymes, Alcalase and Neutrase, it is demonstrated
that even a short hydrolysis increases the solubility to a higher
level than that of the starting material - Purina 500 E. There
seem to be no gross differences between the solubilities obtained
with the two different enzymes when compared at the same DH-value.

At DH-values higher than 3%, it appears that the solubility
curves are virtually flat. This surprising result may be explained
by the following assumptions: 1) The hydrolysis does presumably
not occur as a uniform and simultaneous degradation of all the
protein molecules. Rather, it may follow the one-by-one reaction
whereby the hydrolysate, at any time, contains fairly small peptides and unconverted protein molecules (1). This is a hypothesis
which is still to be investigated. 2) The subsequent acid treatment denatures the remaining high molecular weight material, possibly by disulphide interchange. This denatured material is fairly
insoluble at all pH values as demonstrated by the control sample.
3) The low molecular weight material in the hydrolysate is unlikely to form aggregates and precipitate at the isoelectric
point.

The flat solubility curves can thus be explained by assuming
that the hydrolysates are two-phase systems. One phase is denatured protein, which is nearly insoluble, and the other phase
consists of small, highly soluble peptides. The proportion between the amounts of the two phases is determined by the DH-value
and is reflected in the varying levels of the solubility curves.

The gel chromatograms shown in Figure 6 seem to substantiate
the hypothesis above. The hydrolysates apparently contain both
high molecular weight material and small peptides, but only limited amounts of intermediate peptides.

The results above may be characteristic for these particular
protein-enzyme systems. Thus, Beuchat (23) found that the minimum in solubility at pH 4 for peanut protein is maintained after
a short hydrolysis using e.g. pepsin, bromelain or trypsin.

Emulsifying Capacity. The emulsifying capacity of the modified proteins is improved significantly compared to the unmodified control sample (Figure 7). Owing to the denaturation of the
protein by the acid inactivation, the control sample has a lower
emulsifying capacity than Purina 500 E. It also appears from Figure 7 that the emulsifying capacity of the modified protein can
be increased by at least approximately two times by treatment
with Alcalase to a DH-value of about 5%. The effect of treating
with Neutrase is not as large as with Alcalase and this may be
ascribed to the different specificity of the two enzymes, as reflected in their different hydrolysis curves (Figure 1).

Later experiments (unpublished data) seem to confirm the existence of the maximum in emulsifying capacity described above, although the emulsifying capacities achieved on the individual hydrolysates exhibited a certain variation (±15%, calculated from

Relative peak height
at 206 nm

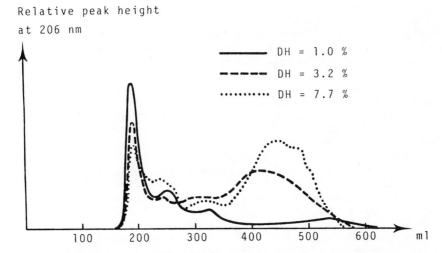

Figure 6. *Gelchromatography of soy protein hydrolysates on Sephadex G 50.*
The hydrolyses were carried out using alcalase at pH 8.0 as described in the
experimental section.

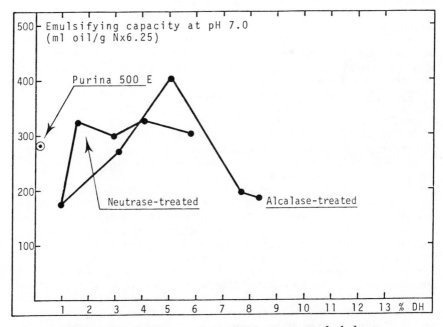

Figure 7. *Emulsifying capacity vs. DH for soy protein hydrolysates*

6 hydrolysates, each with a DH-value of approx. 5%).

A comparison of the curves of Figure 7 with those of Figures 4 and 5 clearly shows that the solubility and the emulsifying capacity are not correlated. The optimal conditions for emulsification seem to be at a DH-value where the hydrolysate consists of approximately equal amounts of soluble and insoluble material. Emulsification involves both hydrophilic and hydrophobic groups in the same molecules and it is therefore important that the molecules are not too small. Also, film formation and surface denaturation play a role, and this also implies that the molecules should not be too small. On the other hand, a certain solubility seems to be necessary for achieving the maximum emulsifying capacity.

Similar results as ours have recently been published by Zakaria and McFeeters (24). They studied the emulsifying activity of soy protein hydrolysates made by peptic hydrolysis and observed that with an increasing concentration of free amino groups in the hydrolysate the emulsifying activity exhibited an increase followed by a decrease. However, because of the differences in experimental conditions between their work and ours, a quantitative comparison would not be justified.

Whipping Expansion. Figure 8 shows that the whipping properties are very much improved after treatment with Alcalase. The general picture resembles that of emulsifying capacity which is not unexpected because the mechanism behind the two phenomena is more or less the same. In the case of whipping, the air is considered as the hydrophobic medium.

The improvement of the whipping properties of enzymatically modified soy proteins and casein has already been of use in the baking industry. For example, Gunther (25) has patented a method for producing these products by pepsin hydrolysis.

Viscosity. For homogeneous solution of polymers coiled at random, the viscosity uniformly is increasing with the chain length.

The viscosities of 27 different gelatin hydrolysates (produced for other purposes than the present investigations) are shown in Figure 9 and seem in general to decrease gradually with increasing DH, as expected. The scatter can presumably be attributed mainly to the uncertainties of the DH determination (0.3 - 0.5% absolute) and to the fact that a DH-value gives the number-average instead of the viscosity-average peptide chain length.

The picture is somewhat different if we consider an inhomogenous system such as soy protein hydrolysate. Figure 10 shows that the viscosity as a function of DH is characterized by a sudden, initial large drop between DH = 1% (control sample) and the hydrolysates with the lowest DH-values. No further drop in viscosity is observed when DH is 3% or above. The figure also shows that the final level of viscosity seems to be dependent on the enzyme rather than on DH which is in marked contrast to the results obtained with gelatin hydrolysates (Figure 9).

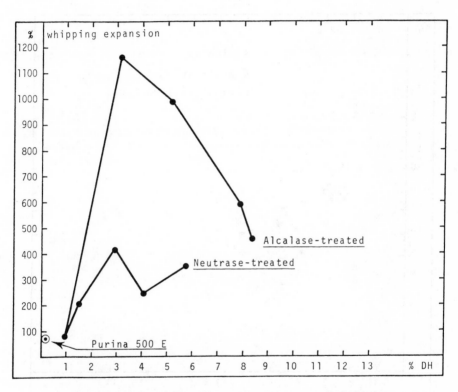

Figure 8.   Whipping expansion vs. DH for soy protein hydrolysates

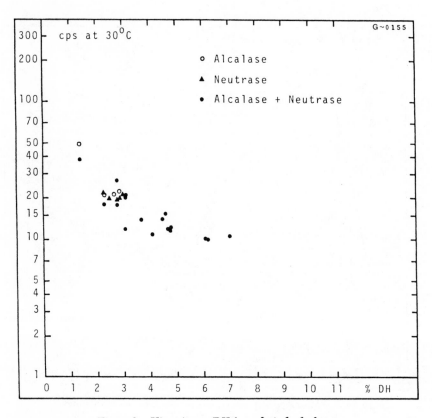

Figure 9.   Viscosity vs. DH for gelatin hydrolysates

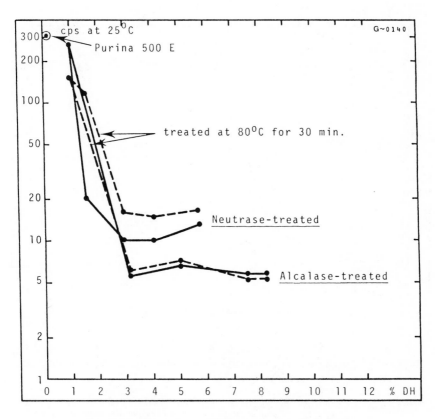

Figure 10.   Viscosity vs. DH for soy protein hydrolysates

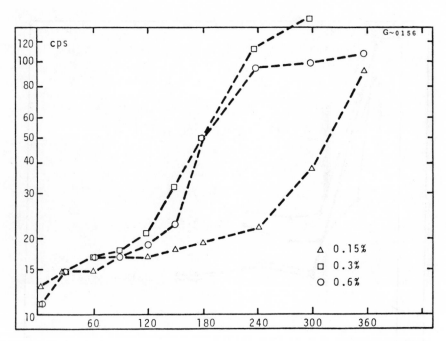

Figure 11.   Modification of corn gluten with neutrase 0.5 L (5); viscosity as a function of hydrolysis time.

Concentration of substrate (S): 8.0% protein (N × 6.25). Concentration of enzyme (E/S): 0.15, 0.3, and 0.6%, neutrase 0.5 L. pH (initial): 7.0. Temperature: 50°C. The viscosity was measured in a rotational viscosimeter at a constant shear rate of 441 sec^{-1}.

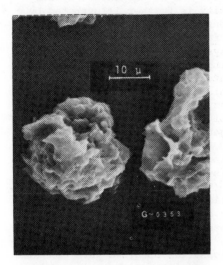 

*Figure 12. Scanning electron micrographs of modified corn gluten (~ 1095 ×).
Left picture shows the product obtained after 300 minutes treatment with 0.3% neutrase
at pH 7 and 50°C (see Figure 10, product denoted with small squares). The right picture shows a control treated under identical conditions except that no enzyme was added
(the viscosity of the control was below 15 cps, see Figure 10). The particle size was very
varied in both samples, and at a lower magnification no difference in the distribution of
particle size could be observed. (The scanning electron micrographs were taken at
Teknologisk Institut, DK-2630 Taastrup, Denmark.)*

The heat treatment of the Alcalase-treated hydrolysates does not cause any changes in the viscosity, in contrast to the Neutrase-treated samples which show an irreversible increase in viscosity, in particular at DH = 1.5%. No gel formation was observed in these experiments, whereas Pour-el and Swenson (26) were able to introduce gel formation capacity in soy protein hydrolysates. This indicates that the removal of the iso-electric soluble phase as carried out in their work (26), is necessary for obtaining gelation ability. Pour-el also proposed that the presence of small peptides would hinder the three-dimensional cross-linking necessary for gel formation (27).

The difference between homogeneous and heterogeneous hydrolysates with respect to their viscosity is further illustrated by Figure 11. The figure shows that enzymatic hydrolysis of corn gluten causes the viscosity to increase more than ten times. Corn gluten is a highly heterogeneous system, consisting of nearly insoluble particles of grossly varying size. With respect to functionality, the unmodified corn gluten resembles sand, with only a small water holding capacity. This functional property increases several fold as a result of the enzyme treatment. It has been proposed that the effect of the hydrolysis may be ascribed to an increase in exposed sites for hydrogen bonding on the surfaces of the particles (5). This would tend to strengthen the cohesion between the particles. Simple swelling did not seem to be an explanation because the gluten particles looked unchanged when viewed in a scanning electron microscope (Figure 12).

The magnitude of the above-mentioned increase in viscosity depends on the hydrolysis conditions as well as on the particular corn gluten chosen for the experiments; the latter is illustrated by the fact that after a reduction in size (by ball milling) of the corn gluten particles, the viscosities obtained were considerably lower (28). It is therefore to be expected that corn gluten from different sources would yield rather different results.

Conclucion

The experiments described in this paper were designed as tentative investigations on the effect of limited hydrolysis of soy protein isolate and they are by no means intended to be a comprehensive study of the relationship between the peptide chain length and the taste and functionality of soy protein hydrolysates. However, we believe that some interesting possibilities of improving certain functional properties have been demonstrated and we plan to continue this work. In particular, we want to investigate whether it is fruitful or not to link the functional properties to DH. Finally, we would like to point out that the problem of relating the measured functional properties to the actual performance of the protein hydrolysate in the food system, where we have conditions drastically different from the conditions in the model system, is still a subject of much controversy and few successes. Thus, we must conclude that the tailor-made functional protein hydrolysates are still a fairly distant goal.

Acknowledgement

We wish to thank Mrs. Gudrun Poulsen, Mr. Poul Erik Andersen, and Mr. Torben Brandt Sørensen for their skillful assistance.

Abstract

The average (number-average) peptide chain length is inversely proportional to the DH (degree of hydrolysis), which is a principal determinant for the taste and the functional properties of protein hydrolysates. The intensity of bitterness and DH are positively correlated at low DH-values, and the relationship is generally expected to show a maximum at a medium DH-value. This may be explained by the generally acknowledged theories concerning the role of the hydrophobic side-chains and the structure of peptides in solution. A series of hydrolysates were prepared using different enzymes and different DH-values and some of their functional properties were studied. Even at a low DH-value the solubility curve is markedly changed, in particular around the iso-electric pH, and the solubility curves are notably flat over a pH-range of 2 - 8. The level increases, as expected, with DH. The emulsification capacity reaches a maximum at a medium DH and decreases at higher DH-values. The foaming capacity shows a similar behaviour albeit the maximum lies at a lower DH-value. The viscosity measurements are discussed in comparison with viscosity measurements on hydrolysates of gelatin and corn gluten, thereby illustrating the differences between homogeneous and heterogeneous systems in this respect. The visocsty of heterogeneous systems may exhibit an increase rather than a decrease in viscosity with increasing DH-value.

Literature Cited

1. Adler-Nissen, J. J. Agric Food Chem. (1976) 24, 1090-93.
2. Richardson, T., In Feeney, R.E., and Whitaker, J.R., "Food Proteins, Improvement through Chemical and Enzymatic Modification", Adv. Chem Ser. (1977), 160, 185-243.
3. Ricks, E., Ridling, B., Iacobucci, G.A. Myers, D.V., in Adler-Nissen, J. Eggum, B.O. Munck, L., Olsen, H.S., "Biochemical Aspects of New Protein Food", FEBS 11th Meeting, Copenhagen, 1977, Vol. 44, pp. 119-128, Pergamon Press, Oxford, 1978.
4. Clegg, K.M., McMillan, A.D., J. Food Technol. (1974) 9, 21-29.
5. Adler-Nissen, J., Process Biochem. (1977) 12 (6), 18-23, 32.
6. Adler-Nissen, J., Ann. Nutr. Alim. (1978) 32, 205-216.
7. Novo Industri A/S, "Proteolytic Enzymes for the Modification of Food Proteins", IB 163, Bagsvaerd, 1978.
8. Jacobsen, C.F., Léonis, J., Linderstrøm-Lang, K., Ottesen, M., in Glick, D., "Methods of Biochemical Analysis", Vol. IV, pp. 187-202, Interscience Publishers Inc., New York, 1957.
9. Novo Industri A/S, "Hydrolysis of Food Proteins in the Laboratory", IB 102, Bagsvaerd, 1978.

10. Steinhardt, J., Beychock, S., in Neurath, H., "The Proteins", Vol. II (2nd ed.), pp. 160-171, Academic Press, New York & London, 1964.

11. Adler-Nissen, J., "Determination of the Degree of Hydrolysis of Food Protein Hydrolysates by Trinitrobenzenesulphonic Acid" (1978) (submitted for publication).

12. Moroney, M.J., "Facts from Figures", 3rd ed., pp. 336-339, Penguin Books Ltd., Harmondsworth, 1967.

13. Arai, S., Noguchi, M., Yamashita, M., Kato, H., Fujimaki, M., Agric. Biol. Chem. (1970) 34, 1338-1345.

14. Swift, C.E., Lockett, C., Fryar, A.J., Food Technol. (1961) 15, 468-473.

15. Eldridge, A.C., Hall, P.K., Wolf, W.J., Food Technol. (1963) 17, 1592-1595.

16. Ney, K.H., Z. Lebensm.-Untersuch. Forsch. (1971) 147, 64-71.

17. Clegg, K.M., Lim, C.L., Manson, W., J. Dairy Res. (1974) 41, 283-287.

18. Schalinatus, E., Behnke, U., Nahrung (1975) 19, 447-459.

19. Matoba, T., Hata, T., Agric. Biol. Chem. (1972) 36, 1423-1431.

20. Fujimaki, M., Arai, S., Yamashita, M., in Feeney, R.E. and Whitaker, J.R., "Food Proteins Improvement through Chemical and Enzymatic Modification", Adv. Chem. Ser. (1977) 160, 156-184.

21. Aso, J., Yamashita, M., Arai, S., Fujimaki, M., J. Biochem. (1974) 76, 341-347.

22. Schellman, J.A., Schellman, C., in Neurath, H., "The Proteins", 2nd ed., Vol. II, p. 35, Academic Press, New York & London, 1964.

23. Beuchat, L.R., Lebensm.-Wiss.u.Technol. (1977) 10, 78-83.

24. Zakaria, F., McFeeters, R.F., Lebensm.-Wiss.u.Technol. (1978) 11, 42-44.

25. Gunther, R.C., US Patent 3,814,816 (1974).

26. Pour-el, A., Swenson, T.S., Cereal Chem. (1976) 53, 438-456.

27. Pour-el, A., Personal Communication, 1978.

28. Adler-Nissen, J., US Patent US SN 758776 (1978).

RECEIVED October 20, 1978.

Functional Properties of Microwave-Heated Soybean Proteins

D. L. ARMSTRONG and D. W. STANLEY

Department of Food Science, University of Guelph, Guelph, N1G 2W1, Ontario, Canada

T. J. MAURICE

General Foods Limited, Cobourg, Ontario, Canada

Although the functional properties of proteins are of the greatest concern to the food scientist, our knowledge of them remains empirical. Observations are but rarely related back to fundamental physicochemical properties such as conformation. In many cases it is impossible even to connect one study with another since the researcher is faced with a myriad of techniques used for measuring protein functionality. It is of great importance to gain an understanding of the relationship between structure and functionality so that the latter can be correctly predicted, manipulated and controlled, if only to insure better utilization of protein in a world where population and malnutrition are growing daily.

This study stems from an attempt to find alternatives to the present method of obtaining protein from heated, solvent extracted soy flakes. As the primary process can cause protein denaturation, insolubilization, impairment of flavour, functionality and color and may possibly affect its nutritive value, it would seem that a better approach might be the direct isolation of soy protein using less harsh conditions.

One method to achieve this is the preparation of an aqueous extract of soaked beans (soymilk) from which the protein is subsequently precipitated at the isoelectric point, a method favoured for recovering soy proteins with unimpaired functionality (1). However, if the protein is not exposed to elevated temperatures the lipoxidase systems in the soybeans promote off-flavour in the soymilk, and trypsin inhibitors, which have a retarding effect on growth laboratory animals, will not be destroyed. A technique has been developed for processing soymilk in which the beans are soaked in water and then ground in water at a temperature above 80°C to inactivate the lipoxidase enzymes before they can have a significant effect on flavour (2). Although this eliminates the beany flavour previously associated with soymilk the protein is still exposed to high temperatures which can lead to denaturation and insolubilization. Also, the ability of this procedure to inactivate trypsin inhibitors is doubtful since Hackler et al. (3) estimated that soymilk prepared at ambient temperatures required 120 min at 90°C to reduce trypsin inhibitor activity to 5% of the original level.

We have attempted to employ microwave heating to inactivate the objectionable trypsin inhibitors and lipoxidase systems. This process has the advantages of being easily controlled and having great penetration so that a shorter treatment may be used with a concomitant decrease in peripheral over-heating.

MATERIALS AND METHODS

Soybeans

All soybeans used were from the same lot and variety (Harrowsoy 65). Proximate analyses indicated 37.6% protein (N x 5.50), 20.2% fat, 6.9% moisture and 11.0% crude fiber.

Preparation of Soymilk

A 100 g sample of soybeans was soaked for 12 h in 450 ml of tap water at ambient temperature. This increased the moisture level of the beans to 62.1%. If the beans received no heat treatment they were processed immediately. Drained beans were ground in 1 ℓ of tap water (20°C) in a 1 gal stainless steel Waring Blendor at high speed for 3 min. The resultant slurry was passed through a small household centrifugal separator (juice extractor) which trapped the solid particles in a triple layer of cheesecloth. The soymilk was stored at 0-4°C until needed.

Heating Treatments

Microwave irradiated soybeans were produced after the initial soaking procedure using a Litton microwave oven (Model 550, frequency 2450 MHz, output 1250 watts). The beans were placed in the centre of the oven on a paper plate in a layer about three beans thick. Samples were treated for 30, 60, 90 or 120 sec. Soymilk was then prepared immediately as previously described. The method of Mattick and Hand (2) also was used to prepare soymilk. The blender was preheated with a boiling water rinse and boiling water was added to the drained beans; the minimum temperature of the slurry during grinding was 80°C.

Isolation of Soymilk Proteins

Figure 1, a modification of the method used by Puski and Melnychyn (4), shows the preparative procedure used to isolate an acid-precipitated fraction of soymilk protein (RDP).

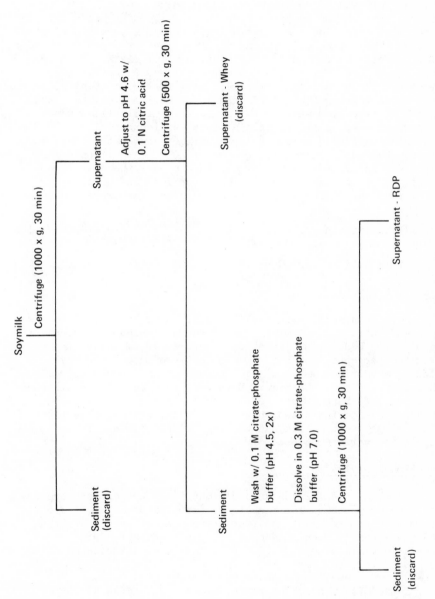

Figure 1.   Scheme for isolation of acid precipitate protein from soymilk, adapted from Ref. 4

Determination of Residual Trypsin Inhibitor

The method of Learmonth (5) as adapted by Van Buren et al.
(6) was followed to determine residual amounts of trypsin inhibitor. This method is based upon the ability of trypsin to
retard gel formation and trypsin inhibitor to suppress the retar-
dation. The source of trypsin was an extract prepared using 30 g
of commercial barley malt flour and 100 ml of distilled water.
This was stirred for 1 h, the mixture allowed to settle over-
night and the supernatant filtered. A buffered gelatin sol was
made by dissolving 8 g gelatin and 1 g disodium hydrogen citrate
in 100 ml distilled water. A mixture of 5 ml distilled water,
5 ml gelatin sol and 5 ml malt extract was incubated at 35°C for
2 h and then placed immediately in an ice water bath. This
control had an average setting time of 20.3 min. Trypsin inhibitor was assayed by using soymilk in place of the distilled
water; the presence of trypsin inhibitor lessens the time for
gel formation. Relative % trypsin inhibitor was expressed as:

$$\text{Relative \% trypsin inhibitor} = 100 - \frac{\text{setting time (soymilk)}}{\text{setting time (water)}} \times 100$$

Six determinations were done in duplicate for each sample.

Sensory Analysis

Soymilk samples were evaluated for odour by a panel composed
of 17-20 untrained members. Panelists were asked to evaluate the
samples for beany odour on an unstructured 10 cm scale ranging
from "no perceptible odour" (0) to "extreme odour" (10) by
placing a vertical mark on the line at a point chosen to reflect
their opinion of the sample. Data were recorded as the distance
from the left hand, 0 end, of the scale to the vertical mark.
The judges also indicated whether the odour was not objectionable, slightly objectionable or very objectionable. Three
replicates were performed.

Functional Analyses

Samples used for analyses of functional properties were the
redissolved protein fractions (RDP) from the unheated, the 60
sec microwave treatment and from the hot water treatments that
had subsequently been freeze-dried, a procedure reported to
result in minimum denaturation (7). At least six determinations
were done for each test.

Solubility. A 1 g sample of protein was added to either
100 ml of 0.3 M citrate-phosphate buffer ranging from pH 2.0 to
8.5 or to 100 ml of pH 7.0 citrate-phosphate buffer extending in
ionic strength from 0.005 to 1.00. The mixtures were blended

for 5 min in a Sorvall Omnimixer at 3000 rpm, centrifuged for 30 min at 1000 x g and 20 ml of the supernatant analyzed for protein.

Emulsion Capacity and Stability. A 0.5 g sample of the freeze-dried protein fraction was redissolved in a minimum of 0.3 M citrate-phosphate buffer at pH 7.0 and mixed thoroughly with 50 ml of 1 M NaCl for 1 min in a Sorvall Omnimixer at 1000 rpm in a one pint Mason jar set in a water bath (20°C). Crisco oil (50 ml) was added to the jar and an emulsion formed by mixing at 500 rpm with simultaneous addition of oil at the rate of 1 ml/min until the emulsion broke. The endpoint was determined by monitoring electrical resistance with an ohmeter. As the emulsion broke a sharp increase (1 KΩ to 35-40 KΩ) was noted. Emulsion capacity was expressed as the total volume of oil required to reach the inversion point per mg protein. This method is similar to that used by Carpenter and Saffle (8) for sausage emulsions. To establish emulsion stability the same procedure was used except that 100 ml of oil was added and a stable emulsion formed by blending at 1000 rpm for 1 min. A 100 ml aliquot was transferred to a graduate cylinder and allowed to stand at room temperature. Observations were made of the volume of the oil, emulsion and water phases at 30, 60, 90 and 180 min.

Bulk Density. Bulk density determinations were made using a Scott paint volumeter (Fisher Scientific Co.) with a 1 in^3 brass receiver to collect the sample.

Foam Volume. The determination of foamability was carried out using the procedure of Hermansson *et al.* (9) with modifications. A 1 g sample of the freeze-dried protein fraction was homogenized with 90 ml of citrate-phosphate buffer in a Sorvall Omnimixer at 3000 rpm. The resultant foam and liquid were transferred to a 250 ml graduated cylinder and the mixer cup washed with 10 ml of buffer. The cup was drained for 2 min and the cylinder allowed to stand for 30 min at which time the foam volume was measured. The influence of pH and ionic strength on foam volume was established using the buffer systems previously described.

Wettability. Wetting time was determined by dropping 1 g of sample from a height of 10 cm onto the surface of a 0.3 M citrate-phosphate buffer in a beaker 7 cm in diameter and measuring the time required for all the sample to wet, as evidenced by a complete color change. The influence of pH was established using the buffer system previously described.

Water Swelling, Water Binding and Dispersion Indices. The methods used were modifications of those employed by Rasekh (10). A 1 g sample was added to 20 ml distilled water in a preweighed, glass stoppered graduated cylinder, mixed with a glass rod and

shaken vigorously for 3 min. The mixture was allowed to stand
for 1 h at room temperature and the foam layer washed down with
5 ml distilled water followed by a further hour of standing. At
this point the dispersion index could be determined by removing a
5 ml aliquot from the midpoint of the suspension and drying it at
105°C for 18 h, its value being the percentage dry weight of the
total sample weight. Alternatively, the supernatant was separated from the sediment by decantation, the volume of the sediment
recorded and the weight of the sediment determined before and
after drying at 105°C for 15 h. The parameters were calculated
as:

$$\text{Water swelling index} = \frac{\text{Swelled volume}}{\text{Precipitate weight}}$$

$$\text{Water binding index} = \frac{\text{Weight of water in sediment}}{\text{Precipitate weight}}$$

Scanning Electron Microscopy

Representative freeze-dried samples were mounted on alumi-
num stubs using conductive paint, coated with gold/palladium
(60:40) in a Technics sputter coater and examined in an ETEC
Autoscan scanning electron microscope at 10 kV.

Electrophoresis

Samples were electrophoresed using a 7% polyacrylamide
system (stacks at pH 8.9, runs at pH 9.5). Sample loading was
between 5 and 15 µl. A current of 4 ma/gel was applied with the
cathode in the upper bath of a Model 1200 Canalco system until
the bromphenol blue tracking dye had moved 3.8 cm through the
running gel. The resulting gels were stained for 1 h with
aniline blue-black (0.5% in 7% acetic acid), destained with 7%
acetic acid in a Model 1801 Canalco quick gel destainer and
subsequently scanned at 620 nm with a Joyce-Loebl Chromoscan
densitometer.

Differential Scanning Calorimetry

A DuPont Model 990 thermal analyzer equipped with a Model
910 DSC cell base was used for differential scanning calorimetry.
Samples were analyzed as 15% (w/w) solutions of freeze-dried RDP
which had been dialyzed to remove excess buffer salts. A heating
rate of 5°C/min was used; runs were performed in a nitrogen
atmosphere (54 psi). A known weight of water was used in the
reference pan to balance the heat capacity of the sample pan.
The instrumental sensitivity was 0.005 (mcal/sec)/in. Heats
of transition (ΔH) were calculated as calories/g protein:

$$\Delta H = \frac{A}{MC} \left(60 \; BE\Delta qs \right)$$

where A = peak area, M = sample mass, C = sample concentration,
B = time base setting, E = cell calibration coefficient,
Δqs = Y axis range.

The transition temperature was taken as the temperature at
the peak maximum.

RESULTS AND DISCUSSION

Yield

Protein recoveries were established for soymilk, RDP and
"whey" fractions as a percentage of the protein in the intact
beans. Table I gives the comparative recoveries for the six
treatments. It may be seen that increasing the microwave treatment significantly reduced the protein yield in soymilk and RDP.
These results paralleled previous findings which indicated that
the degree to which soy proteins can be dissolved at neutral pH
depends upon the extent of heating during processing (11).
Although more protein was lost in the grinding step after microwave processing than the hot water treatment, the corresponding
RDP's gave a higher protein yield; twice as much protein was
recovered from the 60 sec microwave irradiated sample as from
that treated with hot water. Thus, more protein was acid precipitated using moderate microwave heating while hot water
grinding produced proteins that were initially water soluble
but could not be precipitated at pH 4.6. The latter have
probably been heat denatured to the point that they easily
form insoluble aggregates and were removed during the centrifugation steps.

Residual Trypsin Inhibitor Levels

Table II gives the results of residual trypsin inhibitor
levels for the various soymilk preparations. The 90 and 120 sec
microwave treatments were the most effective in inactivating the
trypsin inhibitor complex while hot water treated and unheated
samples showed the highest levels. It is not surprising to find
that microwave processing is more efficient than hot water in
suppressing trypsin inhibitor considering the rapid penetration
of food material by microwaves and the susceptibility of protein
action to small heat induced changes in tertiary structure.
Hence, Collins and McCarty (12) found microwaves produced a more
rapid destruction of endogenous potato enzymes (polyphenol oxi-
dase and peroxidase) than hot water heating.
It is difficult to judge a safe level of trypsin inhibitor
but Van Buren et al. (6) have shown that maximum protein efficiency ratios are obtained when at least 90% of the trypsin

Table I. Influence of heating methods on protein recovery.

Treatment	Protein recovery (% of protein in whole bean)		
	Milk	RDP	Whey
Unheated	77.7^c	59.7^d	9.1^{b,c}
30 sec microwave	72.6^c	46.3^c	7.6^{a,b}
60 sec microwave	41.3^b	21.8^b	7.6^{a,b}
90 sec microwave	34.8^a	11.3^a	10.3^c
120 sec microwave	30.7^a	8.6^a	11.2^c
Hot water	71.4^c	10.7^a	5.1^a

a,b,c,d Columns bearing similar superscripts do not differ significantly (P $\leq$ 0.05).

Table II. Influence of heating methods on residual trypsin inhibitor levels.

Treatment	Relative trypsin inhibitor level (%)
Unheated	94.6^d
30 sec microwave	54.9^b
60 sec microwave	13.5^a
90 sec microwave	0
120 sec microwave	0
Hot water	73.5^c

a,b,c,d Numbers bearing similar superscripts do not differ significantly (P $\leq$ 0.05).

inhibitor has been destroyed with conventional heating of soymilk at 93°C; this means heating times of 60 min (3), much longer than the present process.

Although the soybeans used in this study were soaked prior to microwave treatment, this may not be necessary. It has been observed (Pour-El, personal communication) that irradiation of soybeans containing only innate moisture (6-7%) for a period comparable to those used in this work reduced trypsin inhibitor levels by 90%. Allowing the microenvironmental water of the protein to be the energy transmitter reduced the time needed for inactivation. It was postulated that adding moisture actually reduced the process efficiency because of the energy required to heat the additional water.

Sensory Analysis

Results of the sensory analyses of the various soymilk samples are shown in Table III. Judging from the percent favorable ratings (not objectionable plus slightly objectionable responses) and the odour ratings, the hot water treatment was highly effective in destroying the lipoxidase enzyme systems. Considerable microwave energy is needed to approach this effect and even with a 120 sec treatment a higher odour rating was obtained. However, consideration of sensory analysis, trypsin inhibitor levels and recovery data led to the choice of the 60 sec treatment to be compared with hot water processing for further study.

It will be noted that the two proteins, trypsin inhibitor and lipoxidase, are apparently influenced differently by the microwave treatment. While microwave heating is more effective than hot water in destroying trypsin inhibitors (Table II) the reverse is true for lipoxidases (Table III).

Functional Properties

For these analyses the freeze-dried RDP samples used all had protein contents of approximately 50%. Calculations are given on a constant protein basis.

Solubility. Figure 2 shows the solubility curves for the three heat treatments as a function of pH and ionic strength. The usual solubility minima occurred in the isoelectric region but the microwave treated and the unheated proteins had the narrowest pH range of insolubility (Table IV). This table also gives solubilities of pH 6.0 and 8.0 and shows the higher solubility of the microwave treated protein at the former pH. This was somewhat unexpected since heat processing generally leads to insolubilization but it may be that microwave irradiation produces a form of microdenaturation unaccompanied by a loss in solubility.

It appears that ionic strengths of 0.025 to 0.5 are optimal

Table III. Influence of heating methods on soymilk odor.

Treatment	Odor rating	% Favorable ratings
Unheated	7.44^d	33^a
30 sec microwave	4.88^c	64^b
60 sec microwave	$3.76^{b,c}$	78^c
90 sec microwave	$2.97^{a,b}$	89^d
120 sec microwave	2.36^a	96^d
Hot water	1.53^a	94^d

a,b,c,d Numbers bearing similar superscripts do not differ
significantly ($P \leq 0.05$).

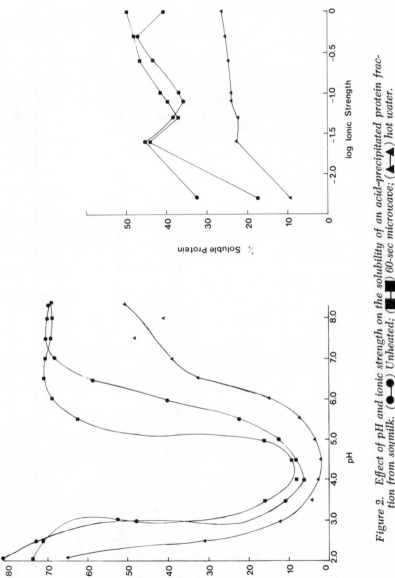

Figure 2. *Effect of pH and ionic strength on the solubility of an acid-precipitated protein fraction from soymilk. (●—●) Unheated; (■—■) 60-sec microwave; (▲—▲) hot water.*

Table IV. Influence of heating methods on functional properties of soy protein.

Property	Heating method		
	Unheated	60 sec microwave	Hot water
Range of less than 20% solubility (pH units)	2.4[a]	1.7[a]	3.5[b]
% Solubility at pH 6.0	39.6[b]	68.9[c]	15.2[a]
8.0	69.1[b]	68.8[b]	40.3[a]
Emulsion capacity (ml oil/mg protein)	2.60[c]	2.11[b]	1.55[a]
Emulsion stability (% emulsion phase)	31[b]	22[b]	4[a]
Foam volume (pH 7.0, ml)	24.6[a]	21.1[a]	23.9[a]
Bulk density (g/in^3)	5.00[b]	5.19[b]	3.72[a]
Wettability (pH 7.0, min)	41.6[c]	15.0[b]	5.2[a]
Water swelling index	5.03[a]	5.80[b]	7.01[c]
Water binding index	3.10[a]	3.91[b]	5.83[c]
Dispersion index	2.43[a]	2.53[b]	6.28[c]

a,b,c Rows bearing similar superscripts do not differ significantly ($P \leq 0.05$).

for maximum solubility and the 60 sec microwave treated and the unheated samples were more soluble at all ionic strengths than those of the hot water treatment (Figure 2). Generally, it has been reported that reduced solubility of plant proteins is associated with lower ionic strength (13, 14). This phenomenon has alternatively been explained as a consequence of dissociation of the protein into subunits due to strong intramolecular repulsive forces (13) or the failure of low ionic strength solutions to supply the net charge required for protein dispersibility (15).

Emulsion Capacity and Stability. The emulsion capacities of the three protein preparations are shown in Table IV and again both the unheated and microwave treated samples were significantly superior to that of the hot water process. Although it is difficult to relate these data to those of other workers, since a wide variety of conditions are employed for measuring this property, it has generally been found that emulsifying properties are related to the aqueous solubility of proteins (7) which further bears on the capacity of proteins to lower interfacial tension between hydrophobic and hydrophilic components.

Figure 3 gives the results of emulsion stability determinations. It was found that the proportions observed after 3 h remained constant for a further 2 days. These data substantiate those found for emulsion capacity in that hot water grinding significantly reduces the ability to soy proteins to form a stable emulsion.

Foam Volume. Moving in either direction from the isoelectric point caused a considerable increase in foam volume (Figure 4) but this is especially evident for the unheated and microwave treated samples. Increasing ionic strength seemed to favor increased foam volume. This particular functional property, along with many others, is difficult to interpret because of the present state of confusion existing in the literature. For example, in the review of protein functional properties by Kinsella (7) reference is made to studies that have shown maximum foaming of proteins close to the isoelectric point as well as proteins whose foam volumes appeared independent of pH. Ionic strength is reported to have varying effects on foaming and, while a certain degree of heating is advantageous to this property, higher temperatures resulted in impaired foaming. Other important factors in foaming such as the presence of bound lipids, degree of denaturation, oxidation/reduction conditions and, of course, test conditions have been mentioned, all of which make it very hard to relate foam volume to other functional properties.

Wettability. All samples required more time to wet close to the isoelectric point and increasing the degree of heating appeared to favour shorter wetting times (Figure 5). Since

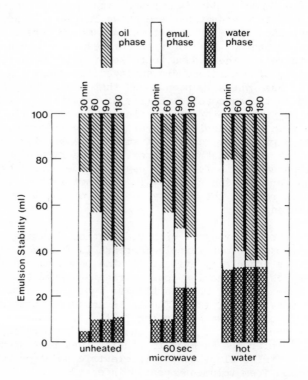

Figure 3. Effect of heat processing method on emulsion stability of an acid-precipitated protein fraction from soymilk

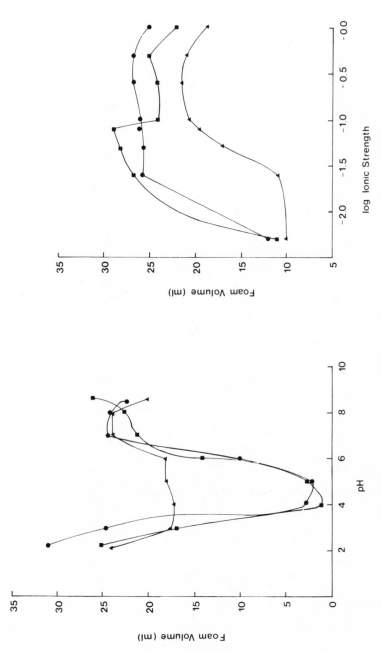

Figure 4. Effect of pH and ionic strength on the foam volume of an acid-precipitated protein fraction from soymilk. (●—●) Unheated; (■—■) 60 sec microwave; (▲—▲) hot water.

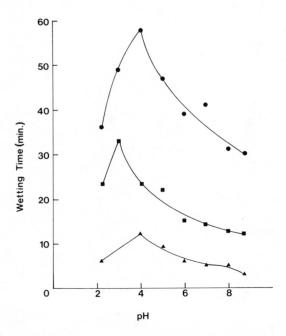

*Figure 5.   Effect of pH on the wettability of an acid–precipitated protein fraction
from soymilk*

ionic charge apparently influences wettability it could be that
denaturation leads to an unfolded or elongated conformation
exposing more charged groups to interact with water molecules.
This, however, may not be as simple as it appears; Rasekh (10)
has suggested that the physical nature of the protein can
influence wetting time and Table IV shows some correlation
between wettability and bulk density. Also, all samples have
an appreciable amount of non-protein components which can influence both these parameters.

Water Swelling and Water Binding and Dispersion Indices.
Table IV shows water swelling, binding and dispersion indices
for the three soy proteins fractions. Hermansson (13) reported
that heating soy isolate produced an increase in swelling, the
largest effect being at 80°C. She also indicated that a highly
soluble protein has poor water binding properties, a relationship seen with these heated soymilk proteins. As with wettability, electrostatic attractions seem important and denaturation
favorably influences the property. Thus, functional property
indices related to water/protein interactions could reflect the
conformation of polypeptide chains within the protein.

Scanning Electron Microscopy

Micrographs of the freeze-dried RDP preparations are shown
in Figure 6. The unheated and microwave treated samples are
clearly differentiated from those treated with hot water. The
former consist of ragged fragments containing numerous but small
pores while the latter appeared more aggregated and exhibits
larger orifices. A consideration of bulk density (Table IV) and
microstructure may help to explain some aspects of protein/water
interaction properties. Porosity and particle size could be
important parameters, however they are difficult to control and
are rarely measured in studies of functional properties.
    These data agree with those (16) describing an isoelectric
precipitate of soybean protein as having a globular structure
while prior heat denaturation yielded a rough surface with pores
of different sizes. It was also reported that water holding
capacity was much higher for the heated protein. The authors
concluded that a heat treatment was necessary to form the three-
dimensional network characteristic of the aggregated material and
that the microstructure of protein curd was related to physical
and textural properties since the heated curd was more springy
and cohesive than nonheated curd.

Electrophoresis

The major protein fractions in the samples were identified
and named according to Cumming et al. (17) although two peaks
(D, Rm = 0.60-0.70; E, Rm = 0.70-0.78) appeared as a result of

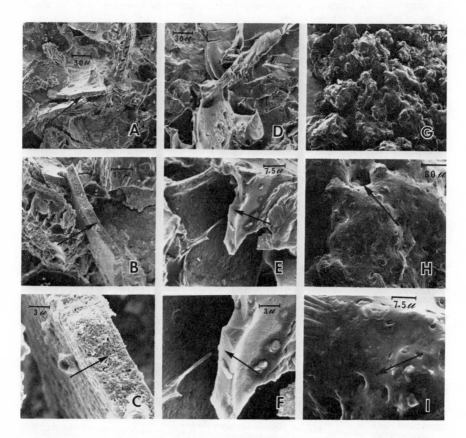

Figure 6.   Scanning electron micrographs of freeze-dried protein fractions from soymilk: (a, b, c) untreated, (d, e, f) 60-sec microwave, (g, h, i) hot water treatment. Note relative size of pores (arrows). (●—●) Unheated; (■—■) 60-sec microwave; (▲—▲) hot water.

processing that are not found in the Catsimpoolas (18) scheme. Table V presents data on the effect of heat processing on the relative distribution of protein in soluble fractions of soymilk and RDP; sketches of the densitometer tracings for these samples are shown in Figure 7. The patterns that emerge indicate that, as heat processing increases in severity, higher molecular weight moieties dissociate to produce smaller fractions which may then go on to form insoluble aggregates. The lower solubility of the hot water treated material may be due to this breakdown and aggregation. The apparently anomalous higher solubility of the irradiated sample relative to the unheated one cannot be explained by degree of heat treatment only as will be seen in the DSC data. Rather, the two proteins seem to have a different quantitative composition in that the more soluble protein has a higher relative amount of the 11S component. The reason for this is unknown.

Recently, a study was made of the electrophoretic as well as other physical and chemical properties of three types of soybean protein fractions (19). It was found that heating generally reduced the solubility of a suspension of these fractions and also increased their viscosity. These changes were attributed to subunit dissociation and aggregation.

Differential Scanning Calorimetry

Calculated heats of transition for the RDP fractions are shown in Figure 8. These data can be used as a measure of the nativeness of a protein but this should be done cautiously in proteins which are, in fact, mixtures of proteins that exhibit different ΔH values. In such a case the heat of transition observed for the whole sample will be an average value dependent upon the proportions of constituent proteins present.

For example, in this work it was observed that the proportion of the 7S (apparent transition temperature - 77°C) and 11S (apparent transition temperature - 91°C) fractions varied with microwaving time and these two fractions probably have different ΔH values. In this case the 7S fraction decreased proportionately greater than the 11S fraction as a function of heating time. Thus, if differential fractionation occurs in an isolation step, it may be said that a drop in ΔH reflects in part a loss of nativeness. Decreases in ΔH may be seen to parallel increasing microwave treatment and also reflect changes in protein properties such as yield, trypsin inhibitor level, color rating and certain functional properties. Thus, this method should prove to be an important tool for studying functional properties through the use of nonempirical thermodynamic data.

Table V. Influence of heating methods on relative distribution of protein fractions.

Treatment	Fraction (% of total area)					
	A (15S)	B (11S)	C (7S)	D	E	F (2S)
Unheated milk	19.8[a]	23.4[b]	9.9[b]	-	-	5.2[a]
RDP	20.1[b]	28.2[a]	-	-	-	5.4[a]
60 sec microwave milk	15.9[a]	27.7[b]	7.0[a]	5.0[a]	-	9.8[b]
RDP	17.5[b]	35.4[b]	11.5[a]	3.7[a]	-	6.2[a]
Hot water milk	19.7[a]	10.0[a]	11.5[b]	12.0[b]	4.4	8.9[b]
RDP	10.9[a]	25.1[a]	11.7[a]	5.2[b]	-	10.0[b]

a,b Columns bearing similar superscripts do not differ significantly (P $\leq$ 0.05).

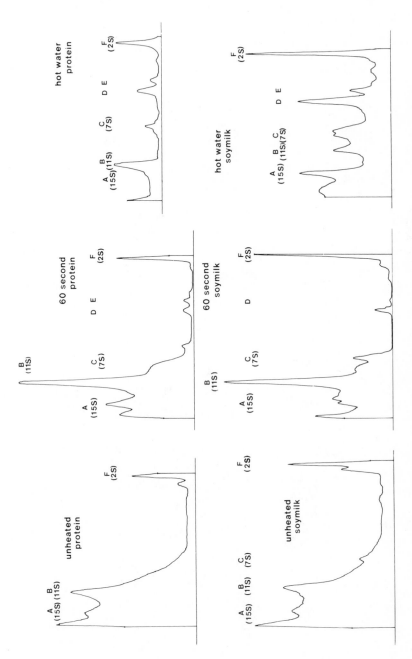

Figure 7. Densitometer tracings for soymilk proteins and RDP fractions as a function of heat processing method

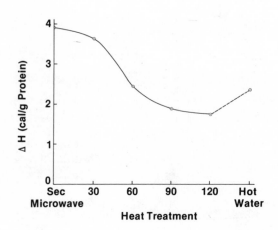

Figure 8.   Heats of transition for RDP fractions as a function of heat processing method

SUMMARY AND CONCLUSIONS

Heat processing, no matter what mode, can have a denaturing effect and thus influence functional properties of soy proteins. Yet, heat is necessary in order to inactivate the trypsin inhibitors and the lipoxidase enzyme systems responsible for antinutritional effects and beany odours of products such as soymilk. If all parameters monitored in this study are considered simultaneously it may be concluded that an initial microwave treatment of soaked beans for 60 sec produces a protein product superior to conventional hot water processing methods. While with this treatment there was a decrease in amount of soluble protein compared to unheated samples, it gave higher yields of acid precipitable protein than hot water grinding. When functional properties are considered, the microwave method produced proteins that were consistently superior to the standard method, and, while usually slightly inferior to unheated beans, this difference was often not statistically significant. A notable exception was solubility; irradiated proteins were more soluble than unheated ones over an important pH range. Microwave treated beans produced soymilk having an acceptable trypsin inhibitor level and tolerable odour rating as well.

The major effect of heat, as evidenced by electrophoretic patterns of soymilk and an acid precipitated protein fraction, was a breakdown of higher molecular weight fractions into lower molecular weight moieties. Microwave treatment produced a different electrophoretic protein profile, one higher in the 11S component than did the hot water process. This may explain the higher solubility of the former protein. Functional properties were also affected; increasing heat treatment produced a protein that had poorer solubility, a wider isoelectric range, less emulsion capacity and stability but greater water swelling, water binding and dispersion indices and better wettability.

Finally, it may be possible to speculate as to the general influence of heat on the functional properties of soy proteins. Figure 9 gives some functional properties and the influence on them of increasing temperature. Results of studies on the major storage proteins of soybean (1) tell us that they exist in a random or β-pleated sheet configuration and the relatively large numbers of nonpolar amino acids cannot be all buried in the interior of the molecule. Also, the proteins are large and a subunit structure is indicated.

It may be reasonable, as indicated in Figure 9, to consider, under appropriate pH and ionic strength conditions, a native molecule formed of associated subunits bonded by hydrophobic, ionic and hydrogen bonds, probably in that order of importance. Heating this structure could dissociate the subunits but their mutual attraction would be expected to lead quickly to aggregation into larger but still soluble complexes. Further heating may cause the subunits to unfold. Disulfide cleavage would

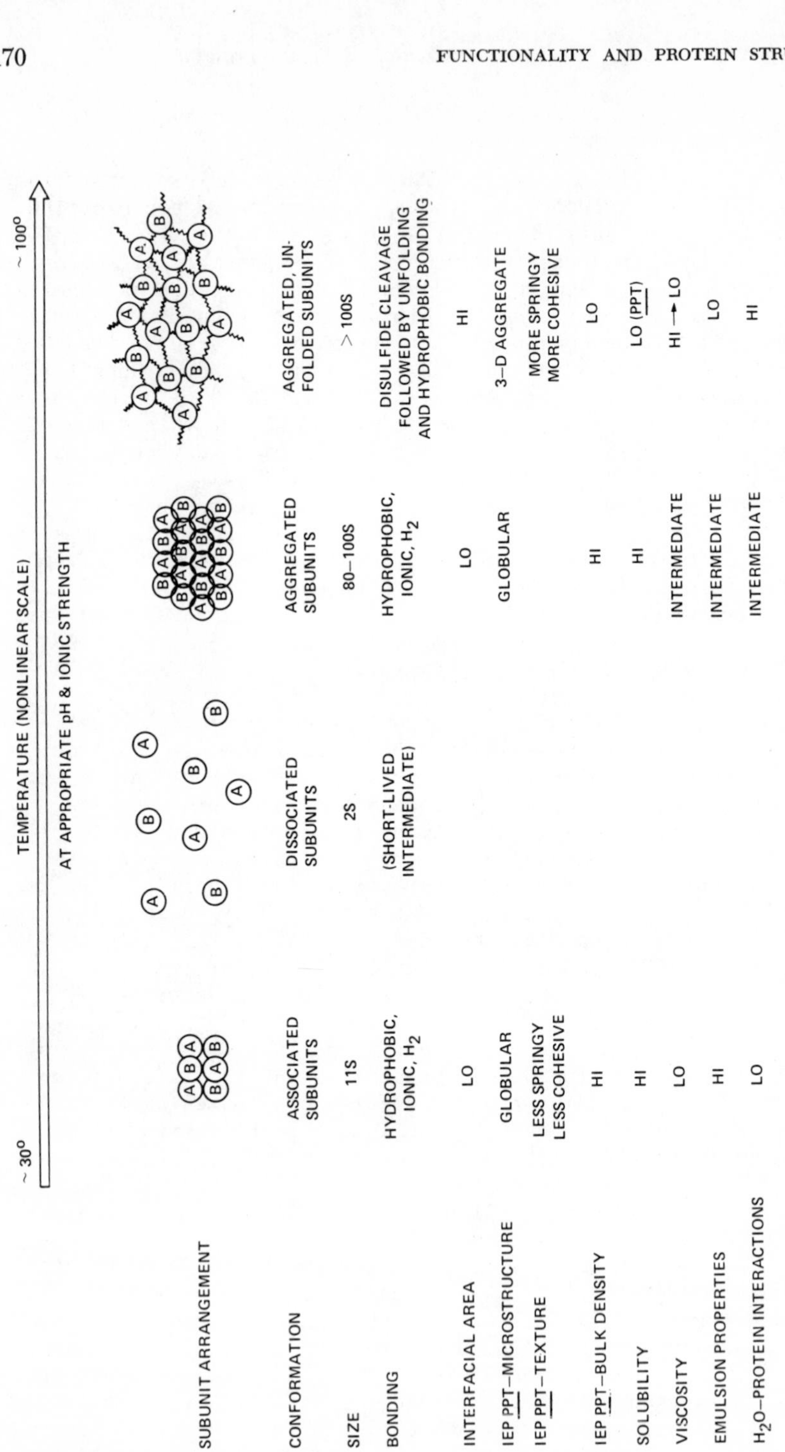

TEMPERATURE (NONLINEAR SCALE)   ~30° → ~100°

AT APPROPRIATE pH & IONIC STRENGTH

SUBUNIT ARRANGEMENT	(associated)	(dissociated)	(aggregated)	(aggregated, unfolded)
CONFORMATION	ASSOCIATED SUBUNITS	DISSOCIATED SUBUNITS	AGGREGATED SUBUNITS	AGGREGATED, UN-FOLDED SUBUNITS
SIZE	11S	2S	80–100S	> 100S
BONDING	HYDROPHOBIC, IONIC, H_2	(SHORT-LIVED INTERMEDIATE)	HYDROPHOBIC, IONIC, H_2	DISULFIDE CLEAVAGE FOLLOWED BY UNFOLDING AND HYDROPHOBIC BONDING
INTERFACIAL AREA	LO		LO	HI
IEP PPT–MICROSTRUCTURE	GLOBULAR		GLOBULAR	3-D AGGREGATE
IEP PPT–TEXTURE	LESS SPRINGY LESS COHESIVE			MORE SPRINGY MORE COHESIVE
IEP PPT–BULK DENSITY	HI		HI	LO
SOLUBILITY	HI		HI	LO (PPT)
VISCOSITY	LO		INTERMEDIATE	HI → LO
EMULSION PROPERTIES	HI		INTERMEDIATE	LO
H_2O–PROTEIN INTERACTIONS	LO		INTERMEDIATE	HI

Figure 9. A possible structural explanation for the influence of heat on the functional properties of soybean proteins

expose more hydrophobic groups and lead to even larger and insoluble aggregates. The loss of electrostatic repulsion between electrical double layers of like sign could occur at this point, concomitant with a loss of bound water, and make London-Van der Waal's forces important. The interfacial area of the protein would increase as unfolding proceeds.

Unfolding of native protein subunits would be expected to be accompanied by a sharp drop in ΔH but the dissociation of subunits may not demand unfolding. Thus, while Privalov and Khechinashvili (20) showed for five single chain proteins that ΔH varied directly with denaturation temperature, our data (Figure 8) exhibits enough lack of linearity to question the conclusion that with a large, subunited protein only an unfolding phenomenon is being observed. It may be that the initial part of the curve represents subunits dissociating. Obviously, further work on the fundamental aspects of thermal denaturation in this soybean system is needed.

ACKNOWLEDGEMENTS

This work was sponsored in part by the National Research Council, the Department of Industry, Trade and Commerce and the Ontario Ministry of Agriculture and Food. Technical assistance was provided by Mrs. C. Burgess, Mrs. B. Holmes and Miss P. Pierson.

ABSTRACT

Aqueous extracts of soybeans were prepared from beans that had been either microwave processed or ground in hot water and protein fractions were acid precipitated to establish the effects of the processing methods on functional properties. Soymilk proteins which had received no heat treatment retained optimal functional properties (with the exception of foaming and those properties relating to the interaction of proteins with water) and gave the highest yield of protein. Residual trypsin inhibitor levels and objectionable odours were, however, highest in this preparation. Proteins from soymilk prepared by the hot water method exhibited poor isoelectric precipitation and functional properties, a low level of objectionable odour but high levels of trypsin inhibitor. An initial microwave exposure gave a superior product compared to the conventional process and higher protein solubility at certain pH's than even the unheated sample. Electrophoresis, scanning electron microscopy and differential scanning calorimetry were used to suggest that these observations resulted from the action of heat in first dissociating protein subunits followed by their aggregation and unfolding.

LITERATURE CITED

1. Smith, A.K. and Circle, S.J.   "Soybeans: Chemistry and
    Technology", Smith, A.K. and Circle, S.J., Eds., AVI
    Publishing Co., Westport, Conn., 1972.
2. Mattick, L.R. and Hand, D.B., J. Agric. Food Chem. (1969)
    17, 15.
3. Hackler, L.R., Van Buren, J.P., Steinkraus, K.H., El Rawi,
    I. and Hand, D.B., J. Food Sci. (1965) 30, 723.
4. Puski, G. and Melnychyn, P., Cereal Chem. (1968) 45, 192.
5. Learmonth, E.M., J. Sci. Food Agric. (1952) 3, 54.
6. Van Buren, J.P., Steinkraus, K.H., Hackler, L.R., El Rawi,
    I. and Hand, D.B., J. Agric. Food Chem. (1964) 12, 524.
7. Kinsella, J.E., Crit. Rev. Food Sci. Nutr. (1976) 7, 219.
8. Carpenter, J.A. and Saffle, R.L., J. Food Sci. (1964) 29,
    774.
9. Hermansson, A.M., Sivik, B. and Skjöldebrand, C., Lebensm.
    Wiss. Technol. (1971) 4, 201.
10. Rasekh, J., J. Milk Food Technol. (1974) 37, 78.
11. Wolf, W.J. and Tamura, T., Cereal Chem. (1969) 46, 331.
12. Collins, J.L. and McCarthy, I.E., Food Technol. (1969) 23,
    337.
13. Hermansson, A.M., "Proteins in Human Nutrition", Porter,
    J.W.G., and Rolls, B.A., Eds., Academic Press, London,
    1973.
14. Hermansson, A.M., Olsson, D. and Holmberg, B., Lebensm. Wiss.
    Technol. (1974) 7, 176.
15. Nash, A.M., Kwolek, W.F. and Wolf, W.J., Cereal Chem. (1971)
    48, 360.
16. Lee, C.H. and Rha, C., J. Food Sci. (1978) 43, 79.
17. Cumming, D.B., Stanley, D.W. and deMan, J.M., J. Food Sci.
    (1973) 38, 320.
18. Catsimpoolas, N., Cereal Chem. (1969) 46, 369.
19. Bau, H.M., Poullain, B., Beaufrand, M.J. and Debry, G.,
    J. Food Sci. (1978) 43, 106.
20. Privalov, P.L. and Kechinashvili, N.N., J. Mol. Biol. (1974)
    86, 665.

RECEIVED October 17, 1978.

Using Some Physicochemical Properties of Proteins in Coacervate Systems

J. K. SORENSEN, T. RICHARDSON, and D. B. LUND

Department of Food Science, University of Wisconsin—Madison, Madison, WI 53706

Coacervates have been extensively investigated for their possible role as protocells in the origin of life on earth. Early work in this area has been reviewed by Yevreinova (1). Other research on coacervation has centered around microencapsulation of substances to isolate or selectively release them for industrial applications (2-10). The purpose of this paper is to review the formation of coacervate systems and the characteristics and alterations of those characteristics of coacervates, and to discuss the results of our investigations of coacervates as models for plant tissue cells. Much of the background material discussed in this paper is based on the excellent review of coacervates by Yevreinova (1).

Homogeneous, transparent solutions of proteins, carbohydrates, and other compounds can separate into two layers, one depleted and one enriched with these compounds. The process of separation of macromolecules into discrete entities is termed coacervation. The layer rich in molecules of the dissolved substance, referred to as the coacervate layer, actually consists of liquid "drops" or spherical microcapsules. The equilibrium liquid, which is the medium adjoining the coacervate layer, always contains less substance than the original solutions. The discrete liquid droplets resulting from macromolecular interactions might be made to serve as pseudocells from which pseudo tissues might be derived to constitute a restructured food.

Coacervates and Their Formation

Classification and Nomenclature. Coacervates can be divided into simple ones and complex ones based on the complexity of their chemical composition. Simple coacervates form when a compound with a great affinity for water is added to a solution of a hydrophilic molecule, causing its dehydration and a decrease in its solubility. Molecules of the same chemical composition are thus involved in simple coacervation. Complex coacervates are obtained when solutions of positively charged molecules and negatively

0-8412-0478-0/79/47-092-173$05.00/0

charged ones are mixed. The molecules interact and their charges
are neutralized.

Forces Affecting Coacervate Formation. The existence of
coacervates also depends on the ratio of the forces of attraction
and repulsion between the molecules involved in coacervation.
Since coacervates are composed of hydrophilic, high molecular
weight compounds, hydrogen bonds are readily formed. Van der
Waals forces and polar forces which form a surface charge and
then a double layer of ions on molecules and micelles greatly
affect coacervates. Electrostatic forces cause attraction between opposite charges, which leads to their interaction and a
decreased free energy of the system. Forces of repulsion which
exist include electrostatic forces between like charges and
hydration forces.

Surface and Aggregation Phenomena. The surface tension
arising at the interface between the coacervate drops and the
equilibrium liquid also determines the existence of coacervate
drops. An insufficient surface tension will result in the dis-
solution of the drops. The magnitude of the surface tension
depends on the proportion of the substances making up the coacervate and also on extraneous compounds added to the system as
shown in Figure 1. According to Yevreinova (1), certain coacervate drops or layers have a surface tension of 0.0015 to 2.31
dynes/cm.
Coacervate drops are usually assumed to be formed from solutions. However, they can also be formed from precipitates of
organic compounds. Yevreinova (1) observed, using a microscope,
a gradual conversion of precipitates into coacervate drops.
Aqueous solutions (0.67%) of gum arabic and gelatin were combined
in the ratio of 3:5 and acidified with 4% acetic acid to pH 3.5-
4.0. The acidification caused the separation of a flocculent
precipitate which changed into drops on heating to 40°C, over the
course of 30-40 minutes.
Consequently, when coacervates are formed from precipitates,
a dilution of the precipitates takes place, the precipitates
swell and produce drops. If the drops are formed from solutions,
a concentration of molecules in the drops is observed.
Coacervate drops may be converted into a layer or a flocculent precipitate on standing, but they may be reconverted into
drops and solutions. This reversibility may have definite limits,
because it is related to the conditions under which the coacervate was formed and its chemical composition. For instance, if
the composition of a coacervate includes serum albumin, the
number of conversions is usually limited to three or four, after
which it is irreversibly denatured.

pH Effects. The pH range for the formation of complex coacervate is bounded by the isoelectric points of the molecules of which they are composed, which allows the molecules to be oppositely charged. As the pH of the system becomes closer to the isoelectric point of one of the substances than the other, correspondingly more of that substance is needed for the neutralization of the charges and for the coacervate to form. The relationship between the pH and the ratio of concentration of gum arabic to gelatin for a complex coacervate composed of gelatin and gum arabic is shown in Figure 2. Also, as coacervates are made by combining the components in varying ratios, the pH of such coacervates will change correspondingly.

The pH can also determine the type or composition of the coacervates formed from a three-component mixture, such as gum arabic, gelatin, and yeast nucleic acid, for example. In such a coacervate, at pH 3.36, the drops consist mainly of gum arabic and gelatin; at pH 3.48, of gelatin, gum arabic, and nucleic acid; at pH 3.8, of gelatin and nucleic acid.

Because the isoelectric points of alkaline proteins are generally in the range of pH 7 or above, coacervates whose composition includes alkaline proteins can be obtained at a more alkaline pH than when acid proteins are used. The greater the difference between the isoelectric points of the macromolecules, the more readily do they form coacervates.

Characteristics of Coacervates

Chemical Nature. Coacervate systems have been formed from numerous protein-protein, protein-carbohydrate, protein-phosphatide, protein-lipid, protein-nucleic acid, and multicomponent mixtures. The coacervate drops, or microcapsules, have been investigated for encapsulating flavor oils, colorants, enzymes, flavor enhancers, leavening agents, microorganisms, vitamins, minerals, amino acids, sugars, salts, etc. for use in foods (4). They have also been used in adhesives and reactive resins, graphic products, drugs, household specialty items, perfumes, agricultural chemicals (insecticides, fertilizers, etc.), and many other products (8,9,10).

Electronic Charge Effects. Alterations in chemical and physical variables of the coacervate system change its characteristics. By characterization of the coacervate systems, and then manipulation of these variables, their action can be oriented in the desired direction.

Measurement of the cataphoresis rates of the solutions at various pH values will aid in obtaining coacervates. The higher the absolute product of the cataphoresis rates of the solutions, the more readily coacervates are formed, and the optimum pH value necessary for coacervation can be found from the value of the cataphoresis rate. The cataphoresis rate changes as a

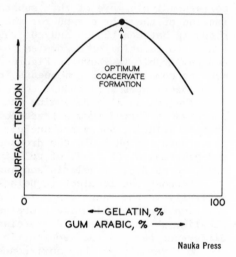

Nauka Press

Figure 1. Surface tension between co-
acerbate layer and equilibrium liquid in
protein–carbohydrate coacervate. The
initial concentrations of gelatin and gum
arabic are taken as 100% (1).

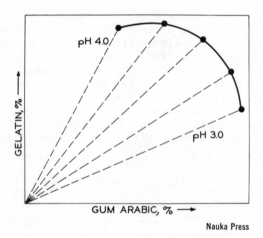

Nauka Press

Figure 2. Formation of coacervates at various pH values as a function of the
ratio of protein and carbohydrate solutions (1)

function of the initial concentration of the solutions from which
the coacervate was obtained. For example, when 1% solutions of
gum arabic and gelatin are used to obtain gum arabic-gelatin
coacervates, the cataphoresis rate is 473, and when 4% solutions
are employed, the cataphoresis rate is 295 (given in arbitrary
units). The rate and direction of the motion of the coacervate
drops toward the positive or negative pole in cataphoresis may
change with the magnitude and sign of the charge on the coacervate. The charge of the coacervate may change if different proportions of the oppositely charged molecules from which the
coacervate is to be obtained are employed or if the pH of the
medium changes. Isoelectric focusing of macromolecules should
be invaluable in establishing pI values for the macromolecules
and optimum pH values for coacervate formation.

  Viscosity. Since coacervates are heterogeneous liquid systems with a nonhomogeneous distribution of substances, the viscosity of a coacervate layer differs markedly from that of an
equilibrium liquid. The viscosity of the equilibrium liquid is
lower than the viscosity of the solutions from which the coacervate was obtained, whereas the viscosity of the coacervate layer
or drops is higher than that of the initial solutions. The decrease of the viscosity of the equilibrium liquid results from
the decrease in the total volume of the particles, since they
concentrate into larger coacervate drops. Coacervates are formed
most completely when the viscosity of the equilibrium liquid is
lowest.
  The viscosity is influenced by the pH of the medium and the
concentration and relative amounts of the initial solutions from
which the coacervate is formed. The relationship between these
factors is illustrated in Figure 3. The formation of coacervates
is shown as a function of viscosity and concentrations of gelatin
and gum arabic solutions at various pH values.

  Size, Concentration and Organization of the Molecules.
Coacervate drops can range in size from several tenths of a micron
to several hundred microns or even larger in diameter. The size
of coacervate drops is affected by several factors. At the optimum temperature for the existence of a coacervate, the drops
are largest. Coacervate drops are smaller when they consist of
compounds whose isoelectric points are widely separated, and
larger when the compounds have closer isoelectric points. The
diameter of drops also increases with the concentration of the
initial solutions from which the coacervates were obtained (11).
  For each coacervate system, a definite optimum concentration of the initial solutions exists at which the mean diameter
of the drops is 5-8 μ and their number is maximum and equal to
7 to 14 million per ml (11). An increase in the concentration
of the initial solutions leads to the appearance of flocculates
and reduction in the number of drops through sharp acceleration
of their coalescence, while a decrease in the concentration as

compared with the optimum is accompanied by the appearance of a
large number of small drops and a decrease in the total number of
drops (11). The optimum concentrations of initial solutions are
frequently between 0.5% and 2.0%. For gum arabic-gelatin coacervates, the highest initial concentrations at which coacervates
can be obtained are 6%, whereas the lowest concentrations are
0.0016%.

The volume occupied by the coacervate drops is generally
3 to 5% of the volume of the entire coacervate system. An increase in the concentration of the solutions from which the
coacervate is formed leads to an increase of the total volume
occupied by the coacervate layer. For example, when 1% solutions
of protein and carbohydrate were used to form coacervates, the
volume of the coacervate layer was 5.31% of the total volume of
the coacervate system, whereas when 4% solutions of these substances were employed, 26.2% of the volume was occupied by the
coacervate drops. A negative temperature coefficient is characteristic of coacervates whose composition includes fats and
fatty acids: the higher the temperature, the smaller the volume
occupied by the coacervate layer.

Molecules of chemical compounds in the solutions collect in
the coacervate drops. Coacervates are characterized by an increase in the concentration of the dry matter in the drops by a
factor of several tens of times, as compared to the original
solutions, and even more as compared to the equilibrium liquid.
For example, under optimum conditions, gum arabic-gelatin
coacervate drops contain 84% of the molecules, and 16% remains in
the equilibrium liquid. In the coacervate from histone-DNA,
96.3% of all the compounds were contained in the drops. The concentration of the initial solutions from which the coacervates
were obtained was about 0.1%, but the concentration of the substances in the drops increased to 10%.

There is an inverse proportion between the size of the drop
and its concentration of molecules. This phenomenon may be partly because the smaller the radius of the drop, the larger the
compressive force expelling mainly water out of the drop and thus
increasing the concentration of other substances.

It is not known whether a definite spatial ordered arrangement of molecules takes place in coacervate drops. There have
been very few studies which have investigated the molecular
organization of coacervates. It has been shown that coacervation
is associated with a transformation of native protein molecules
into aggregated ones, and it is thought that most of the molecules in coacervate drops or layers may be present in the form
of unstructured aggregates. According to Yevreinova (1), the
relationship between the amount of native and aggregated protein
molecules can change, since the concentration of the substances
in the drops varies considerably with their size and chemical
composition, and native protein molecules appear to be converted
into aggregates as the concentration of molecules in solution
increases.

There is some evidence to indicate an ordered arrangement of molecules in coacervate drops. By measuring the elasto-viscous properties of a protein coacervate layer, Pchelin and Solomchenko (12) observed an ordered arrangement of molecules in the coacervate layer. The birefringence of a coacervate system also suggests a structural state of the molecules, because birefringence is observed in systems where the molecules are oriented in a definite fashion. Solutions of low concentrations of chemical substances, which do not show birefringence, are used to prepare coacervates, but once the coacervates are formed, the coacervate layers show flow birefringence.

Electron micrographs of coacervate drops have shown that some drops appear to have very complex internal structures (13). Evreinova, et al. have observed that coacervate drops have surface film-like layers of different structures either with projections or without them (13). Coacervate drops may contain vacuoles where polymer concentration is greatly reduced.

The ability of the coacervate microcapsule to isolate its contents and/or release them selectively is a characteristic which is important to many of the present and potential uses of coacervates.

The wall of the coacervate drop can be made permeable, semi-permeable, or impermeable to diffusion of molecules through the microcapsule wall. The rate of release of the contents of the coacervate drop or intake of molecules from outside the drop depends on the nature of the polymer(s) which make up the wall material, the thickness of the microcapsule wall, the pore width of the wall, the molecular weight of permeating materials, and the degree to which the polymeric wall materials are cross-linked (2).

Salts and Stabilizers Effects. Electrolytes, specifically mineral salts, have a definite effect on coacervates since they carry a charge and are therefore capable of changing the charge of the coacervate. If an added salt has a greater affinity for water than the coacervate, it dehydrates the coacervate drop, and thus breaks it down and converts it to a precipitate. The more hydrated the coacervate, the harder it is for it to hold water and the less salt is required for its precipitation.

Since mineral salts are electrolytes, they can decrease the charge of the coacervate or even eliminate it completely, which eventually causes the coacervate to precipitate. Salts can also increase the charge of the coacervate, thus increasing its stability. Positively charged coacervates are most sensitive to neutral salts. The higher the valence of the salt, the stronger its influence on the coacervate. As would be expected, different amounts of different salts are required in order to give the same amount of precipitation of the coacervates. In the strength of destructive action on coacervates, the anions and cations of the salt can be arranged in the following sequences:

Sequence of cations: Li>Na; Ca>Mg>Sr>Ba

Sequence of anions: $KCl < \dfrac{K_2SO_4}{2} < \dfrac{K_3[Fe(CN)_6]}{3} < \dfrac{K_4[Fe(CN)_6]}{4}$

Salts can not only break down coacervate layers and drops, but also promote their enlargement. The effect of salts on the formation of coacervates from lecithin and carrageen is seen in Figure 4.

Salts can also influence coacervates by causing the pH at which a given coacervate is formed to change. For example, in the absence of salts, the optimum for the formation of a coacervate from egg albumin and gelatin is at pH 4.82, and on the addition of 20 milliequivalents of KCl, it shifts to pH 3.0. Therefore, by adding different amounts of salt, one can obtain coacervates at different pH values.

In addition to changing the pH of the coacervate, the salt concentration may also change the size of the coacervate drops.

ADP and orthophosphate, when added to a coacervate system from histone and gum arabic, stabilize the coacervate drops ($\underline{14}$). The addition of ADP, ADP with Mg^{+2}, or orthophosphate leads to the conservation of the total volume of the coacervate drops, whereas without those additives, the total volume of the coacervate drops decreases very significantly. The nature of the stabilizing action of the phosphates on histone-gum arabic coacervate drops is not clear. Because of the presence of phosphate groups in the system, the stabilization may be associated with the degree of ionization of the phosphate groups and the peculiarities of the distribution of charges in these molecules ($\underline{14}$). Possibly the stabilization of drops by phosphates is associated with the degree of hydration of the phosphate groups of the compounds.

Functionality of Coacervates

Micelles and Pseudocells. The characteristics and properties of coacervate drops have suggested to us the possibility of using them as "pseudocells" to uniquely restructure foods. Our concept is that coacervate "pseudocells" may be built up into simulated tissue systems through interaction between coacervate drops. These simulated tissue-like structures might then be used to restructure macerated fruit or vegetable tissue. The use of proteins to form coacervates for restructuring macerated plant tissue would increase the total dietary protein intake from plant sources since protein would be added to plant tissue systems.

For the formation of coacervates, proteins have been investigated which provide possibilities for induced interaction between the coacervate drops. Table I shows the combinations of macro-

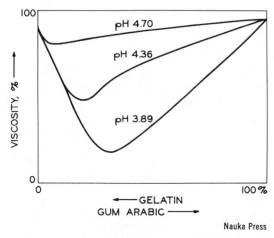

Nauka Press

Figure 3. Viscosity of equilibrium solutions after formation of coacervates from gelatin and gum arabic at various pH values. Initial viscosity and concentrations of gelatin and gum arabic solutions taken as 100% (1).

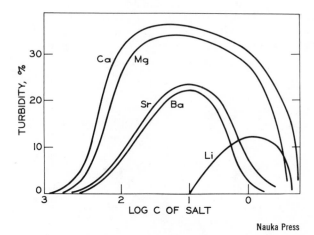

Nauka Press

Figure 4. Effect of salts on the formation of coacervates from lecithin and carrageen. The percentage of turbidity corresponds to the amount of coacervate (1).

Table I

Combinations which did not produce coacervates under
the conditions and concentrations employed in this
study.

Combination	Parts by weight	pH	Observation
gum arabic/alkaline gelatin/ pea puree	1/0.6/134	3.8–4.0	No structure imparted to the puree
kappa-casein/acid gelatin	varied from 1/7 to 7/1	6.5–7.0	No visible evidence
gum arabic/acid gelatin	5/3	6.2	No visible evidence
gum arabic/acid gelatin/ kappa-casein	33/19/1	6.2	No visible evidence
acid gelatin/alkaline gelatin	1/1	6.8	No visible evidence
pectin/gum arabic/ kappa-casein	4/1/3–1/4/3	2.7	No visible evidence
pectin/kappa-casein	5/3	2,3,4,5	No visible evidence
gum arabic/soy protein isolate	5/3	2.5	No visible evidence

molecules under the conditions and concentrations employed in our
studies, which did not produce coacervates. Gum arabic was considered for use in coacervate systems because of the reported
ease with which it forms coacervates with other macromolecules,
and because it has been used extensively in previous investigations. Alkaline gelatin has also been used in many previous investigations. Kappa-casein was considered for coacervates because of the possibility of using enzyme hydrolysis of the kappa-
casein to induce interaction between coacervate drops. Because
of its high (basic) isoelectric point, acid gelatin was considered. It could possibly be used to form coacervates in the
pH range of nonacid foods. Pectin, another polysaccharide with
an isoelectric point similar to that of gum arabic, was considered for use in coacervates because its properties are easily
chemically modified. Soy protein isolate was considered because
it is a readily available commercial protein source with an isoelectric point around 4.6, a pH range appropriate for food systems. Keratin from wool was also of interest for use in coacervates because of its high cysteine and cystine content which
could possibly be utilized to form disulfide bridges between
coacervate drops.

As mentioned previously, coacervates are formed at a pH
between the isoelectric points of the macromolecules of which
they are composed. The desired final pH of the restructured food
determines what the range of the isoelectric points should be,
and therefore what proteins or macromolecules should be used for
the formation of coacervates in the system.

Tissue Simulation. In order to be able to simulate a tissue-
like system, we have attempted to induce an interaction between
coacervate drops by various means. Polyvalent cations were investigated to determine if ionic surface bonds could be formed
between coacervate drops on the addition of polyvalent cations,
causing a texturization of the coacervate drops. The addition
of sodium dodecyl sulfate was investigated to see if interaction
between coacervate drops could be modified. Enzymatic hydrolysis
of the protein contained in the coacervates was also investigated
for inducing an interaction between the coacervate drops. We
also investigated the possibility of incorporating microcrystalline cellulose into the coacervates and subsequently complexing
the coacervates with cations.

Practical Coacervate Formation. The composition of the
various parts of the coacervate system was determined by using
^{14}C-labelled kappa-casein as one component of the coacervates,
and then performing a mass balance on the system. The ^{14}C-kappa-
casein was prepared by reductive methylation of kappa-casein with
^{14}C-formaldehyde (15). This derivatization of kappa-casein was
found not to materially affect its physical properties (16).

Although coacervates could not be formed under some conditions identified above, several systems were developed in which characteristic coacervate drops were formed. Table II lists these systems in which coacervates were formed. Gum arabic-alkaline gelatin coacervates, which have previously been heavily investigated, were formed using 5 parts gum arabic to 3 parts gelatin (w/w), at around pH 4.0. Coacervates were formed from gum arabic and kappa-casein, with 5 parts gum arabic to 3 parts kappa-casein, at around pH 2.7. Photomicrographs of the coacervates are shown in Figure 5. The gum arabic-kappa-casein coacervate drops were decidedly smaller (approximately 1 to 2 μ diameter) than gum arabic-alkaline gelatin coacervates (approximately 7 to 10 μ). Gum arabic-kappa-casein coacervate drops had significantly less surface interaction than gum arabic-alkaline gelatin coacervates, as measured by turbidimetric procedures and shown in Figure 6. Coacervates were also formed from gum arabic, kappa-casein, and alkaline gelatin. The coacervates were formed at approximately pH 3.7, with 5 parts gum arabic to 3 parts protein, where 5, 10, 20, 40, and 60% of the gelatin was replaced, on a weight-to-weight basis, by kappa-casein. It was difficult to conclude, from protein determinations and an attempt to use ^{14}C-labelled kappa-casein, what amount of kappa-casein was actually incorporated into those coacervates. There has also been some evidence of an interaction between keratin and gum arabic, when they were mixed at pH 5.5-6.0 in the ratio 3 parts keratin (w/w) to 5 parts acid gelatin, but further characterization and utilization of keratin for coacervate formation was postponed because of difficulty in preparing keratin. It could be that S-sulfonation of keratin by oxidative sulfitolysis might produce a highly charged species which could be incorporated into coacervates. Subsequent reductive cleavage of the S-sulfonate followed by oxidation of the SH groups might lead to structure formation.

The addition of polyvalent cations (Ca^{++} and Al^{+++}) to a coacervate system was observed to induce surface interactions between the coacervate drops. The amount of surface interactions was evaluated by the rates of settling of the coacervates using visual observation. The maximum interaction was observed on the addition of 4.98×10^{-7} moles Ca^{++} per milligram of protein and 4.98×10^{-8} moles Al^{+++} per milligram of protein. Al^{+++} was about ten times more effective than Ca^{++} in inducing surface interaction between coacervate drops. However, the aggregated coacervates were easily redispersed, suggesting that ionic surface bonding was not sufficient for texturization of the coacervate drops.

To determine the composition of the coacervates, ^{14}C-labelled kappa-casein was incorporated in the coacervates and a mass balance on the coacervate system was performed. The results of these studies are shown in Table III. Sixty percent of the kappa-casein in the gum arabic-kappa-casein coacervate system

Table II

Coacervate systems which were formed
in this study.

Combination	Parts by weight	pH	Observation
gum arabic/alkaline gelatin	5/3	4.0	Generally about 7-10 μ diameter
gum arabic/kappa-casein	5/3	2.7	Generally about 1-2 μ diameter
gum arabic/kappa-casein/ alkaline gelatin	5/0.15/2.85- 5/1.80/1.20	3.7	Difficult to conclude how much kappa-casein is actually in the coacervates

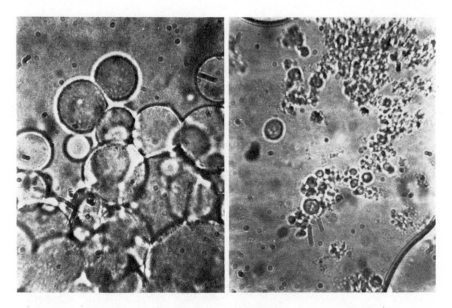

Figure 5.   Photomicrographs of gum arabic/gelatin coacervate (left) *and gum arabic/kappa–casein coacervate* (right), *each at ∼ 470 ×*

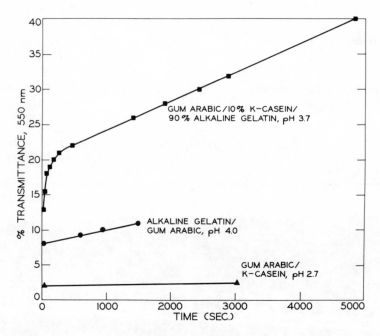

Figure 6.   Comparison of surface interactions of coacervate drops of various co-acervate systems, as measured by rates of settling of coacervate aggregates

Table III

Some characteristics of kappa-casein/gum arabic
coacervates using ^{14}C-labelled kappa-casein.

Treatment	% of the total kappa-casein in the coacervate system[a,b]
Coacervates (centrifuged, washed twice)	59%
Supernatant of coacervates	35%
Washes of coacervates	2%
Peptides released by rennet treatment of coacervates	3%
Peptides released by rennet treatment of control (no coacervates formed, but same quantity of kappa-casein as coacervate system)	36%
Peptides released by treatment of control as with rennet, but deleting the rennet	5%

[a] ± 5%.

[b] Soluble in 2-4% TCA.

Table IV

Stability of coacervates formed from 5 parts
gum arabic/0.3 parts kappa-casein/2.7 parts
alkaline gelatin to pH alterations.

pH	Percent transmittance at 550 nm
2.0	95.5
3.0	46.0
4.0	15.0
5.0	44.0
6.0	89.0
7.0	92.0
8.0	93.0
control, 3.95	12.0

was incorporated into the coacervate drops, and 40% of the kappa-casein remained in the equilibrium liquid.

Gravimetric determinations on coacervate systems have indicated that about 35% of the total macromolecules in the system were incorporated into the coacervate drops. This information, combined with the mass balance data above, indicates that the coacervates were approximately 66% kappa-casein and 34% gum arabic.

Treatment of coacervates with enzymes did not induce an interaction between coacervate drops. By using ^{14}C-labelled protein and performing a mass balance, we found that essentially none of the kappa-casein in gum arabic-kappa-casein coacervates was released by rennet hydrolysis. Pepsin did not appear to hydrolyze the protein in coacervates either. Apparently the proteins are sequestered from enzymic activity as a result of coacervate formation.

Various substances were added to formed coacervate systems to observe their effects. Added sodium dodecyl sulfate did not affect the coacervates. Microcrystalline cellulose particles, added to the coacervate system before and after the coacervates were formed, were observed to be too large to be incorporated in or interact with the coacervate drops. Coacervates made in glucose and sucrose solutions were unaffected by the sugar.

Coacervate drops were quite unstable to pH alterations of the system. The stability of coacervate systems to pH variations was evaluated by measuring the turbidity of systems adjusted to various pH values since turbidity of the mixture serves as an index of the degree of coacervation (17). The coacervate systems formed with gum arabic and protein consisting of 10% kappa-casein and 90% gelatin, at pH 3.8, were most stable at pH 3.8. Some of the coacervates persisted when the pH was varied one pH unit, but were completely dissolved at pH 2.0 and below, and pH 6.0 and above, as shown in Table IV.

Conclusions

In conclusion, coacervates have been formed using food proteins, at pH values feasible for some food systems. We have attempted to induce interactions between coacervate pseudocells by several means, but other alternatives must be developed. Ways of stabilizing the coacervate pseudocells must also be explored, so that more complex structures may be more conveniently built. Several possibilities exist. One is to incorporate egg albumin into the coacervate structure and thermally stabilize the coacervates. Another possibility is that alginate may be incorporated into the coacervate with subsequent stabilization of the pseudocells with Ca^{++}. It may be possible to form a coacervate with alginate and a protein such as α_{s1}-casein, and to treat the coacervate with Ca^{++} to increase its stability. A third possibility is to try to form a phospholipid membrane around preformed coacervate drops. The techniques for forming liposomal particles from phospholipids might be employed. We are continuing to explore coacervates as potential systems for structuring food-related macromolecules.

Acknowledgment

This review was made possible by a grant from the Cooperative State Research Service, United States Department of Agriculture and by support from the College of Agricultural and Life Sciences, University of Wisconsin, Madison, WI 53706.

Literature Cited

1. Yevreinova, T. N. "Concentration of Matter and Action of Enzymes in Coacervates." 222 p., "Nauka" Press, Moscow, 1966.
2. Sliwka, W. Angew. Chem. (1975) 14(8), 539.
3. Vandegaer, Jan E. "Microencapsulation--Processes and Applications." 180 p., Plenum Press, New York, 1973.
4. Bakan, Joseph A. Food Technol. (1973) 27(11), 34-45.
5. Daniels, Roger. "Food Technology Review No. 3: Edible Coatings and Soluble Packaging." 360 p., Noyes Data Corp., Park Ridge, N. J., 1973.
6. Balassa, Leslie L. and Fanger, Gene O. Crit. Rev. Food Tech. (1971) 2(2), 245-265.
7. Sirine, Gloria. Food Product Development. (1968) 2(2), 30-31.
8. Balassa, Leslie L. and Brody, Julius. Food Eng. (1968) 40, 88-91.
9. Flinn, James E. and Nack, Herman. Chem. Eng. (Dec. 4, 1967), 171-178.
10. Balassa, Leslie L. and Weiss, Herbert. Soap and Chem. Special. (1967) 43(12), 163-168.
11. Gladilin, K. L., Orlovskii, A. F., Mamontova, T. V. and Yevreinova, T. N. Biofizika. (1974) 19(2), 259-274.
12. Pchelin, V. A. and Solomchenko, N. Ya. Koloid. Zh. (1960) 22, 63, Cited in Ref. (1), p. 58.
13. Evreinova, T. N., Stefanov, S. B. and Mamontova, T. V. Dokl. Akad. Nauk SSSR. (1973) 208(1), 243-244.
14. Orlovskii, A. F., Gladilin, K. L., Vorontsova, V. Ya., Kirpotin, D. B. and Oparin, A. I. Dokl. Akad. Nauk SSSR (1977) 232(1), 236-239.
15. Means, Gary E. and Feeney, Robert E. Biochemistry. (1968) 7, 2192-2201
16. Zadow, G., Richardson, T. and Olson, N. F. J. Dairy Res. (1978) 45(1), 69-76.
17. Gladilin, K. L., Orlovskii, A. F., Evreinova, T. N. and Oparin, A. I. Dokl. Akad. Nauk SSSR (1972) 206(4), 995-998.
18. Farrell, H. M., Jr. and Thompson, M. P. "Fundamentals of Dairy Chemistry." Byron H. Webb, Arnold H. Johnson, John A. Alford, eds. p. 467, The AVI Publishing Company, Inc., Westport, Conn., 1974.

RECEIVED October 19, 1978.

Structure of Wheat Gluten in Relation to Functionality in Breadmaking

K. KHAN and W. BUSHUK

Department of Plant Science, University of Manitoba, Winnipeg, Canada R3T 2N2

Gluten was first described by the Italian chemist Beccari (1) in 1745. He reported that gluten could be easily prepared by washing the starch and water-soluble components of flour from dough by kneading the dough under a gentle stream of water. The insoluble residue was a viscoelastic mass, later shown to contain about 80% of the total protein of the flour. About two thirds of the mass of gluten is water of hydration. The dry solids contain 75 to 85% protein, depending on the thoroughness of washing, and 5 to 10% lipids. Occluded starch makes up most of the remainder of the dry matter.

Figure 1 shows, by a schematic diagram, how gluten and its major components can be prepared.

Major Components of Gluten

Gliadin. Gliadin is that portion of the gluten proteins that is soluble in 70% aqueous ethanol. It comprises approximately 35 to 40% of the flour proteins. Gliadin imparts the viscous component to the viscoelastic properties of gluten.

Gliadin contains about 50 components identified by a two-dimensional electrofocusing-electrophoresis technique (2). The molecular weights (mol wt) of these components, determined by sodium dodecyl sulfate-polyacrylamide gel electrophoresis (SDS-PAGE), range from approximately 12,000 to 80,000, with the majority of the components having a mol wt of about 36,000 (3). Most gliadin components consist of single chains containing intra-polypeptide disulfide bonds.

The amino acid composition of gliadin (Table 1) shows that approximately 35% of the total amino acid residues consist of glutamic acid. Almost all of the glutamic acid of gliadin is present as glutamine (note high ammonia nitrogen content in Table 1). The high glutamine content promotes hydrogen bonding in the gluten complex. Gliadin also contains a high proportion

0-8412-0478-0/79/47-092-191$05.00/0

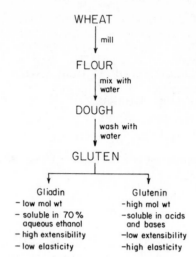

*Figure 1. Schematic of a procedure for
the preparation of gluten and its major
components*

Table I. Amino acid composition[1] of wheat flour components

Amino Acid	Gliadin[2]	Glutenin[2]	Gluten[3]	Flour[4]
Lysine	5	12.5	9	16
Histidine	14.5	13	15	19
Arginine	15	20	20	29
Aspartic acid	20	23	22	33
Threonine	18	26	21	22
Serine	38	50	40	42
Glutamic acid	317	278	290	318
Proline	148	114	137	107
Glycine	25	78	47	27
Alanine	25	34	30	25
Cysteine	10	10	14	18
Valine	43	41	45	37
Methionine	12	12	12	13
Isoleucine	37	28	33	33
Leucine	62	57	59	58
Tytosine	16	25	20	24
Phenylalanine	38	27	32	44
Tryptophan	5	8	6	7
Amide	301	240	298	230

[1]Amino acid residues per 100,000 g
[2]Kasarda et al. (10) (Table 1)
[3]Wu and Dimler (38)
[4]Tkachuk (39)

of proline, approximately 15% of the total amino acid residues.
Proline creates kinks or bends wherever it occurs in a polypeptide
chain, thereby disrupting the regular secondary structure of the
chain. Gliadin also contains low levels of the basic amino acids
lysine, histidine, and arginine, and low levels of free carboxyl
groups, placing the gliadins among the least charged proteins
known.

From the unique amino acid composition of the gliadin
proteins, one would expect a structure that is quite different
from globular proteins. However, optical rotation studies have
shown that the gliadin proteins possess compact tertiary structures similar to those of globular proteins (4,5).

Gliadin appears to influence the loaf-volume (index of
breadmaking quality) potential of a wheat flour in breadmaking
(6). This property of gliadin was demonstrated through protein
fractionation and reconstitution (baking) experiments. However,
the manner in which gliadin reacts with other gluten components
to influence loaf volume is not yet fully understood.

Recent studies of A-gliadin (an α-gliadin component which
tends to aggregate), the most characterized of the gliadin
proteins, has shed some insight on the possible behaviour of
gliadin proteins, and perhaps on gluten proteins in general, in
the breadmaking process. In dilute acid solution, below pH 3
(.001M HCl), A-gliadin exists in its monomeric form with a mol
wt of 31,000. At this low pH, it is in a partially unfolded
configuration. When the pH is raised (to about 5), at very low
ionic strength, it becomes compactly folded. However, when both
the pH and the ionic strength are raised, its monomers form
aggregates with particle weights in the millions (7,8). Figure 2
(9) shows, in schematic form, the various conformational states of
A-gliadin under different conditions. Examination of the
aggregates with the transmission electron microscope (Fig. 3)
revealed a microfibrillar structure which contained intertwined
fibres with diameters of 70 to 80 Å. The formation of the
A-gliadin aggregates is completely reversible and involves only
secondary forces such as hydrogen and ionic bonds, and hydrophobic interactions. This aggregation-disaggregation behaviour
of A-gliadin may be a general phenomenon exhibited by other gluten
proteins (10) and may play a key role in the function of gluten
proteins in dough during the breadmaking process.

Glutenin. According to the classical Osborne (11) definition,
glutenin is that fraction of the gluten proteins that is
insoluble in 70% aqueous ethanol but soluble in dilute acid or
alkali. It comprises approximately 35 to 45% of wheat endosperm
protein. Glutenin imparts the elastic component to the viscoelastic properties of gluten. It is mainly the glutenin that
undergoes extensive changes during dough mixing and the development of optimum rheological properties required for maximization
of the breadmaking potential (loaf volume) of a specific flour

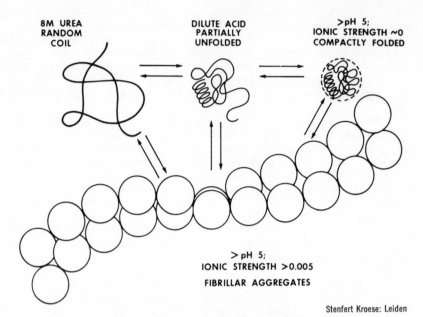

Stenfert Kroese: Leiden

Figure 2. Schematic of conformational structure and aggregate state of A-gliadin under various conditions (9)

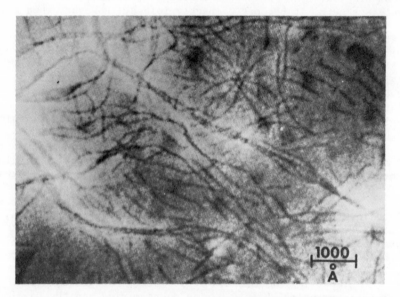

Stenfert Kroese: Leiden

Figure 3. Transmission electron micrograph of A-gliadin microfibrils; preparation negatively stained with uranyl (9)

(12,13).

The key to the functional behaviour of glutenin in bread-making lies in its physical (molecular size and shape) and chemical (amino acid composition, sequence and tendency to aggregate) properties. Gel-filtration (14) and ultra-centrifugation studies (15) revealed that glutenin is a large molecule (or aggregate) with mol wts in the millions. Although the mol wt of glutenin has not been accurately determined, there is considerable evidence which suggests that this protein is extremely poly-disperse and that the mol wt distribution may be a varietal characteristic that is important to breadmaking quality.

Sodium dodecyl sulfate-polyacrylamide gel electrophoresis (SDS-PAGE) analysis of reduced glutenin (Fig. 4) has shown that bread (hexaploid) wheats contain approximately 17 polypeptide subunits, ranging in mol wt from 12,000 to 134,000 (3,16), joined to one another by interpolypeptide disulfide bonds to form long, concatenated structures (17), or by hydrophobic interactions and hydrogen bonds to form highly stable micelles (18). Glutenin of durum (tetraploid) wheats (Fig. 4) lacks three of the largest subunits (134,000, 132,000 and 90,000) present in the glutenin of bread wheats. The poor breadmaking quality of tetraploid wheats has been attributed partly to the lack of all or some of these three high mol wt components of glutenin which presumably play a key role in the function of this protein in dough formation and stability during baking.

Alkylated subunits of glutenin have been fractionated by gel-filtration (19,20,21) into three distinct groups of subunits (Fig. 5). Peak I eluted with the void volume but contained subunits of the lowest mol wt group (68,000 to 12,000 by SDS-PAGE). Peak I subunits appear to exhibit a strong tendency to aggregate, a property which might be extremely important in the functionality of glutenin in breadmaking. Peak II contained the largest subunits (134,000 to 60,000) of glutenin while peak III contained those subunits with the same mobility, by SDS-PAGE, as the two major gliadin proteins (mol wts 35,000 and 45,000). It is presumed that each of the three distinct groups of subunits contributes its unique properties to influence the overall functional properties of glutenin in gluten and in dough.

The amino acid composition of glutenin (Table 1) shows a high content of glutamic acid. Essentially, all of the glutamic acid is present as glutamine. Thus, there are numerous amide groups that can form intra- and inter-molecular hydrogen bonds. This extensive hydrogen bonding is considered to be a very important feature of the physical (rheological) properties of hydrated glutenin. Hydrogen bond-breaking substances (e.g. urea) have a marked influence on the physical properties of dough (22). Glutenin, together with gliadin, are manifestations of the natural selection in the evolution of wheat, which, based on an efficient storage of nitrogen for the new plant, has produced proteins with extremely useful functional properties for

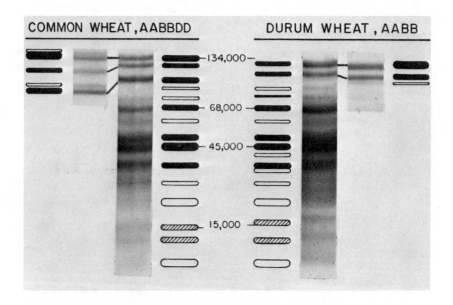

Figure 4.   Typical SDS–PAGE patterns of reduced glutenins of bread and durum wheats

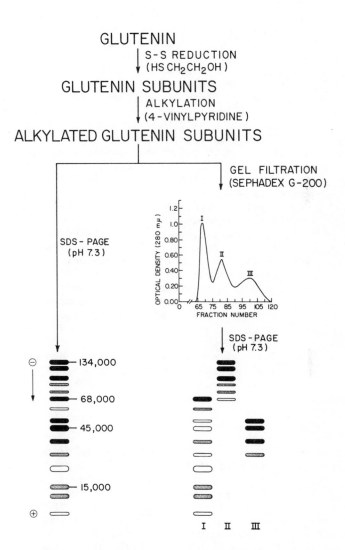

Figure 5.   SDS–PAGE of alkylated glutenin subunits of bread wheat fractionated by gel filtration on Sephadex G-200

breadmaking.

Glutenin contains a relatively high proportion of hydrophobic amino acids such as leucine. The non-polar side chain of leucine can interact with each other, especially in an aqueous environment (as in dough), to form the so-called hydrophobic bonds. A large number of these relatively weak bonds, acting collectively, can contribute substantially to stabilizing glutenin aggregates (23). On the other hand, glutenin contains relatively small proportions of amino acids with acidic or basic side groups, hence its poor solubility in aqueous solvents.

A unique feature of the amino acid composition of the high mol wt sub-units of glutenin (e.g. Peak II of Fig. 5) is the relatively high contents of glycine, proline, glutamine and leucine (Table II). Glycine is present in high proportions in structural proteins such as collagen. As already mentioned, proline disrupts the secondary structure, glutamine promotes hydrogen bonding, while leucine promotes hydrophobic interactions. These four amino acids are probably extremely important in determining the physical structure (hence functionality) of glutenin in gluten and in dough.

Relative to secondary structure, viscosity, sedimentation velocity, ultraviolet difference spectra and optical rotatory dispersion studies (4,24,25) showed that glutenin appears to be an assymetric molecule with a low α-helix content (10-15%). Glutenin contained more α-helix structure in hydrochloric acid solutions and less in urea solutions. The amount of α-helix structure is also influenced by changes in ionic strength (26). The tertiary structure of glutenin is such that it promotes the formation of fibrils under some conditions (27). The secondary, tertiary, and quaternary structures of glutenin can apparently be modified to advantage in the breadmaking process (by oxidizing agents, reducing agents; mechanical development) to produce bread of optimum loaf volume and crumb structure.

Minor Components of Gluten

Lipids. Besides the two major components, the proteins, gliadin and glutenin, gluten also contains substantial amounts of lipids. Lipids form about 5 to 10% by weight of the total solids of gluten. It has been suggested (28) that the function of lipids in gluten is to form a gliadin-lipid-glutenin complex that is essential for the formation of membranes needed for satisfactory gas retention in dough during the breadmaking process. It was also suggested (28) that lipid-binding to gliadin is by hydrophilic (probably hydrogen bonds), and to glutenin by hydrophobic interactions.

Carbohydrates. Carbohydrates (starch, pentosans and sugars) make up most of the remaining (5-15%) solids of gluten. The carbohydrates are thought to modify the properties of gluten during

Table II. Amino acid composition[1] of glutenin and the
protein fractions obtained from gel-filtration of
reduced and alkylated[2] glutenin (see Fig. 5)

Amino acid[3]	Glutenin	Peak I	Peak II	Peak III
Lysine	1.91	3.81	0.73	0.83
Histidine	1.66	2.02	0.73	1.59
Arginine	4.01	4.76	2.14	3.66
Aspartic acid	3.48	7.63	0.69	2.32
Threonine	3.20	4.06	3.03	2.66
Serine	6.26	6.45	6.34	6.49
Glutamic acid	32.24	19.99	40.42	39.32
Proline	12.15	7.93	12.30	14.99
Glycine	9.04	9.75	17.72	3.59
Alanine	4.06	6.96	2.97	2.72
Valine	4.10	5.84	1.58	4.25
Methionine	1.57	1.82	0.87	3.67
Isoleucine	3.01	3.92	0.87	3.67
Leucine	6.79	8.25	4.30	7.34
Tyrosine	3.04	3.41	5.41	1.42
Phenylalanine	3.43	3.39	0.63	4.40

[1]Expressed as mole percent
[2]According to Friedman et al. (40)
[3]Cysteine and tryptophan were not determined

the breadmaking process by diluting the other components (29).
It has been suggested that specific interactions between wheat
starch and other flour components are important to breadmaking
quality; however, the nature of these interactions has not been
identified.

Genetics of Gluten Proteins

The biosynthesis of each polypeptide chain in a living cell
is controlled by a specific gene. The polypeptide chain may be
an individual protein molecule or a portion of a much larger
molecule or functional aggregate. Millions of genes are organized
into particles called chromosomes. The number of chromosomes in
a cell is characteristic of the species.

The chromosome number for bread wheat is 42, comprising 21
pairs of identical chromosomes. The 21 pairs are organized into
three sets of seven chromosomes; each set is called a genome
(Fig. 6). The three genomes of bread wheat are identified by the
letters A, B and D; thus, bread wheat has the genomic composition
AABBDD. On the basis of the number of genomes, which is six,
bread wheats are also called hexaploids. There is considerable
similarity among the three genomes. It has been postulated that
all three originated from the same primitive genome (Fig. 7).
Accordingly, a hexaploid wheat can have as many as three genes
for the same endosperm protein. This multiplicity of genes has
ensured the survival of the wheat plant through the ages, but it
complicates the study of the genetics of its components.

Durum wheat (used for macaroni and spaghetti) is quite
different from bread wheat. Its chromosome number is 28. It has
two of the three genomes of bread wheat and lacks the D genome
(like T. dicoccum in Fig. 7). It also lacks the breadmaking
quality of bread wheats. This is one reason why studies of the
inheritance of breadmaking quality have concentrated on the
chromosomes of the D genome.

Research on the genetics of the gluten proteins has concentrated on the effects of removal of certain genomes, chromosomes
and parts of chromosomes. Wrigley (2) used starch-gel electrophoresis in one dimension and isoelectric focusing in a second
dimension, and located the genes for the synthesis of about 30
of the gliadin proteins. Almost all of the gliadin components
could be assigned to homoeologous chromosomes of group 1 and 6
(30). In a recent genetic study of gliadin, Baker and Bushuk
(31) showed that of the 26 gliadin bands identified by polyacrylamide gel electrophoresis, nine bands were controlled by
single genes and ten of the bands were controlled by two genes.
This study showed also that two groups, one comprising five bands
and the other three bands, were inherited as units and that there
appears to be a linkage between the genes controlling several
bands.

Much of the work on the genetics of gluten proteins related

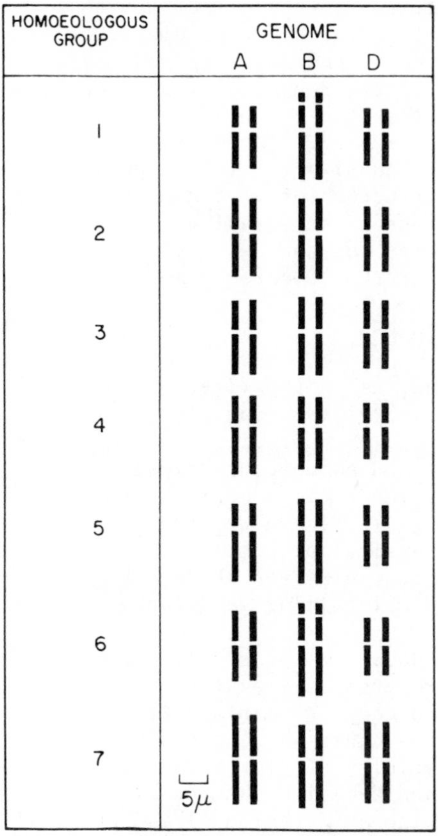

HOMOEOLOGOUS GROUP	GENOME		
	A	B	D
1	‖ ‖	‖ ‖	‖ ‖
2	‖ ‖	‖ ‖	‖ ‖
3	‖ ‖	‖ ‖	‖ ‖
4	‖ ‖	‖ ‖	‖ ‖
5	‖ ‖	‖ ‖	‖ ‖
6	‖ ‖	‖ ‖	‖ ‖
7	‖ ‖	‖ ‖	‖ ‖

5μ

Figure 6. Idiogram of wheat chromosomes based on the results of Gill et al. (41). Durum wheats have the A and B genome chromosomes while common (bread) wheats have the A, B, and D genome chromosomes.

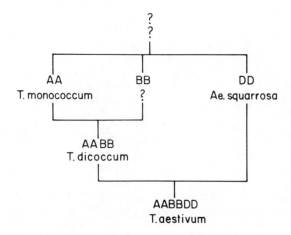

Figure 7. Schematic of the origin of polyploid wheat species

to breadmaking quality has concentrated on the effects on the
glutenin fraction. In one study, the D genome was removed from
four hexaploid varieties to give the so-called extracted AABB
tetraploids (32,33,34). SDS-PAGE of the subunits of reduced
glutenin (Fig. 8) showed that at least two (maybe four) subunits
were deleted when the D genome was removed. The extracted AABB
tetraploids showed poor breadmaking quality (35). It appears
that extensive alteration of the structure of glutenin, due to
loss of key subunits, leads to a loss of breadmaking quality.
It has been shown (36) that the genes for the key glutenin subunits are located on the long arm of the 1D chromosome.

Formation and Function of Gluten in Dough

When flour and water are mixed, as in the first stage of the
breadmaking process, gluten proteins (gliadin and glutenin)
hydrate, interact with each other and with other flour components
(starch, pentosans, lipids, sugars and soluble proteins) to form
a dough. Subsequent mixing (and fermentation) modifies the
viscoelastic properties of the dough in such a way that it
acquires the critical balance in physical properties to retain
the appropriate proportion of the leavening gas while expanding
to produce maximum loaf volume and maintaining satisfactory crumb
structure.
The viscosity of gluten (and dough) is generally attributed
to the gliadin component. This property derives from the
relatively small molecular size and compact tertiary structure
of this component. Glutenin, on the other hand, contributes the
elastic component to the viscoelasticity of gluten (and dough).
This property of glutenin is derived from its chemical (hydrophobic interactions by nonpolar amino acid residues) and physical
structure (high mol wt) of its polypeptide chain. Some polypeptide chains of glutenin may be joined to each other by
interpolypeptide disulfide bonds to form long, compactly folded
concatenations (17). Interpolypeptide disulfide bonds in the
glutenin would withstand a high degree of stretching, hence the
elasticity of glutenin. In addition, elasticity can also result
from the cooperative action of many secondary bonds and interactions that are known to exist in glutenin.
Optimum development of the viscoelasticity of gluten and
dough can be brought about by the physical action of dough
mixing or fermentation (gradual expansion and collapse of gas
cells). Mixing is the most critical stage in the breadmaking
process. It serves to form a three dimensional structure by
transforming particles of gluten proteins into thin membranes
within which are embedded the starch granules and other flour
components. Mixing requires a certain time-limited input of
energy for optimum development of the gluten structure. If the
mixing energy exceeds the optimum level required, then the gluten
membranes are destroyed which is subsequently manifested as a

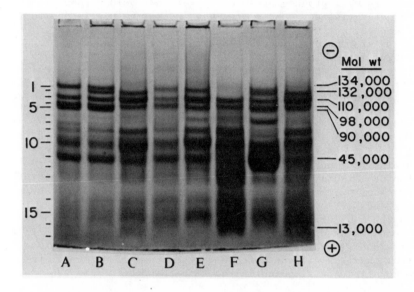

Figure 8. SDS–PAGE patterns of reduced glutenin of hexaploid wheats and their extracted AABB tetraploids: (A) Chinese Spring; (B) Prelude; (C) Tetraprelude; (D) Rescue; (E) Tetrarescue; (F) Stewart 63 (durum); (G) Thatcher; (H) Tetrathatcher (21)

loss of breadmaking quality. The presence of many weak
secondary bonds and interactions facilitates the formation of
optimum gluten structure. However, since the structure so formed
is not permanent, it is susceptible to destruction by overmixing.

The function of the three dimensional structure of dough
developed by mixing is to retain the carbon dioxide produced by
the yeast cells during the fermentation stage of the breadmaking
process while allowing for a certain limited expansion. Since
the gluten proteins possess a characteristic viscoelasticity, its
mambranes can expand to a certain degree. The degree of expansion of the gluten proteins determines the loaf volume of the
particular bread. It is well known that flours of certain wheat
varieties will produce bread of larger loaf volume than flours
of other varieties. The difference in loaf volume is due to a
difference in gluten quality; the flour containing better
quality gluten will expand to a greater degree and will produce
bread of larger loaf volume. The main protein that controls
this expansion appears to be glutenin (12). In baked products
where loaf volume is not an important factor as in cakes, cookies,
crackers and biscuits, a flour of different quality can be used.
Therefore, the type of flour used is determined by the end
product.

Durum wheats, in contrast to bread wheats, form the basis of
the pasta industry. Here a different "quality" of the gluten
proteins is important. The gluten must have an appropriate
ratio of gliadin and glutenin (of appropriate type) for satisfactory flour properties during extrusion (as in spaghetti manufacture) and for the "al dente" bite after cooking.

The many different products that can be made from gluten
proteins demonstrate the versatility of these proteins. With a
better understanding of the structure of the gluten proteins,
cereal chemists will be able to further take advantage of the
enormous industrial potential that these proteins possess.

Summary

Gluten consists of two major proteins, gliadin and glutenin,
the former soluble in 70% aqueous ethanol while the latter is
soluble in acids and bases. Gliadin imparts the viscous
component while glutenin imparts the elastic component to the
viscoelastic properties of gluten and dough. Both proteins are
extremely important in determining the functional properties of
a wheat flour. Glutenin, however, is mainly responsible for the
mixing properties and loaf volume potential of a wheat flour
(37). These two properties, mixing and loaf volume, determine
to a large extent, the final quality of a loaf of bread.

Literature Cited

1. Beccari. Instituto atque academia commentarii (1745), 2 (Part 1): 122.
2. Wrigley, C.W. Biochem. Genet. (1970), 4: 509.
3. Bietz, J.A., and Wall, J.S. Cereal Chem. (1972), 49: 416.
4. Wu, Y.V., and Dimler, R.J. Arch. Biochem. Biophys. (1964), 107: 435.
5. Kasarda, D.D. Baker's Dig. (1970), 44: 20.
6. Hoseney, R.C., Finney, K.F., Shogren, M.D., and Pomeranz, Y. Cereal Chem. (1969), 46: 126.
7. Bernardin, J.E., Kasarda, D.D., and Mecham, D.K. J. Biol. Chem. (1967), 242: 445.
8. Kasarda, D.D., Bernardin, J.E., and Thomas, R.S. Science (1967), 155: 203.
9. Kasarda, D.D., Nimmo, C.C., and Bernardin, J.E. "Coeliac disease. Proc. 2nd Int. Symp.", Stenfert Kroese:Leiden, (1974), 25.
10. Kasarda, D.D., Bernardin, J.E., and Nimmo, C.C. "Advances in Cereal Science and Technology," American Assoc. of Cereal Chemists, Inc., St. Paul, Minn., (1976), 169-171.
11. Osborne, T.B. Carnegie Inst. Wash. Publ. (1907), No. 84.
12. Orth, R.A., Baker, R.J., and Bushuk, W. Can. J. Plant Sci. (1972), 52: 139.
13. Tanaka, K., and Bushuk, W. Cereal Chem. (1973), 50: 590.
14. Bushuk, W., and Wrigley, C.W. Cereal Chem. (1971) 48: 448.
15. Jones, R.W., Babcock, G.E., Taylor, N.W., and Senti, F.R. Arch. Biochem. Biophys. (1961), 94: 483.
16. Khan, K., and Bushuk, W. Cereal Chem. (1976), 53: 566.
17. Ewart, J.A.D. Sci. Fd. Agric. (1977), 28: 191.
18. Kobrehel, K., and Bushuk, W. Cereal Chem. (1977), 54: 833.
19. Huebner, F.R., and Wall, J.S. Cereal Chem. (1974), 51: 228.
20. Arakawa, T., Yoshida, M., Morishita, H., Honda, J., and Yonezawa, D. Agric. Biol. Chem. (1977), 41: 995.
21. Khan, K. "Ph.D Thesis." University of Manitoba, Winnipeg, Canada (1977).
22. Jankiewicz, M., and Pomeranz, Y. Cereal Chem. (1965), 42: 37.
23. Chen, K., Gray, J.C., and Wildman, S.G. Science (1975), 190: 1304.
24. Wu, Y.V., and Cluskey, J.E. Arch. Biochem. Biophys. (1965), 112: 32.
25. Cluskey, J.E., and Wu, Y.V. Cereal Chem. (1966), 43: 119.
26. Wu, Y.V., Cluskey, J.E., and Sexson, K.R. Biochem. Biophys. Acta. (1967), 133: 83.
27. Orth, R.A., Dronzek, B.L., and Bushuk, W. Cereal Chem. (1973), 50: 688.
28. Hoseney, R.C., Finney, K.F., and Pomeranz, Y. Cereal Chem. (1970), 47: 135.
29. Hoseney, R.C., Finney, K.F., Shogren, M.D., and Pomeranz, Y.

Cereal Chem. (1969), 46: 117.
30. Shepherd, K.W. "Proc. 3rd Int. Wheat Genet. Symp.," p. 86.
    Aust. Acad. Sci., Canberra (1968).
31. Baker, R.J., and Bushuk, W. Can. J. Plant Sci. (1978),
    58: 325.
32. Orth, R.A., and Bushuk, W. Cereal Chem. (1973), 50: 680.
33. Bietz, J.A., Shepherd, K.W., and Wall, J.S. Cereal Chem.
    (1975), 52: 513.
34. Khan, K., and Bushuk, W. Cereal Chem. (1977), 54: 588.
35. Kaltsikes, P.J., Evans, L.E., and Bushuk, W. Science (1968),
    159: 211.
36. Orth, R.A., and Bushuk, W. Cereal Chem. (1974), 51: 118.
37. Orth. R.A., and Bushuk, W. Cereal Chem. (1972), 49: 268.
38. Wu, Y.V., and Dimler, R.J. Arch. Biochem. Biophys. (1963),
    102: 230.
39. Tkachuk, R. Cereal Chem. (1966), 43: 207.
40. Friedman, M., Krull, L.H., and Cavins, J.F. J. Biol. Chem.
    (1970), 245: 3868.
41. Gill, B.S., Morris, R., Schmidt, J.W., and Mann, M.S.
    Can. J. Genet. Cytol. (1963), 5: 326.

RECEIVED October 31, 1978.

The Mechanism of Enzyme Protein-Carrier Protein Binding: Physical Chemical Considerations

J. R. GIACIN[1] and S. G. GILBERT

Food Science Department, Cook College, Rutgers University, P. O. Box 231,
New Brunswick, NJ 08903

It has been long recognized that a number of enzymes in a living cell are bound to the membraneous structures of the cell. For example, Green et al.(1) described a pH-dependent binding of several glycolytic enzymes to the erythrocyte membranes and concluded that the entirety of the glycolytic system was associated with the membrane and is not free in solution.

A similar conclusion was arrived at by Arnold and Pette (2) from studies carried out on the in vitro binding of aldolase glyceraldehyde phosphate dehyrogenase, fructose-6-phosphate kinase, phosphoglycerate kinase, pyruvate kinase and lactate dehydrogenase to the structural proteins: F-actin, myosin, actomyosin and stromaprotein.

A number of enzymes appear therefore to be localized in a specific micro-environment, which can influence their biocatalytic activity. Because of the complexity of biological membranes, our understanding of the influence of micro-environmental effects on membrane-bound enzyme is minimal. An important contribution to better understanding the mode of action of membrane-bound enzyme has been the development of the concept of heterogeneous catalysis by enzymes synthetically bound to water-insoluble supports. These immobilized enzymes were viewed as models for the cellular bound enzyme (3,4).

In 1972, Vieth, Gilbert and Wang (5) reported the use of the structural protein collagen as a carrier for the immobilization of enzymes. Subsequent studies by these and other workers have established the general utility of reconstituted collagen as a carrier for enzyme and whole microbial cell binding. (The following references and the references cited therein provide a useful review of the literature on enzyme and whole microbial cell attachment to collagen):(6, 7, 8, 9, 10).

---

[1] Current address: School of Packaging, Michigan State University, East Lansing, MI 48824.

0-8412-0478-0/79/47-092-207$05.00/0

Elucidation of the mechanism involved in collagen-enzyme complex formation (carrier protein-enzyme protein binding) has both theoretical and practical significance. Collagen-bound enzymes can provide a reasonable model for studying interactions between enzymes and cellular membrane and for determining the influence of such a microenvironment on the activity of the enzyme. Enzymes covalently bound to chemically activated collagen films had previously been proposed as simplified models for mitochondrial membrane bound enzymes (11,12). Reactions involving covalent bond formation are not characteristic of a biological system. On the other hand, collagen bound enzyme obtained by complexation (enzyme protein-carrier protein binding) can provide a more reasonable model. In addition, knowledge of the binding mechanism should be useful in the design of more efficient immobilized enzyme reactors for practical application of immobilized enzyme catalysis.

In elucidating the mechanism of collagen-enzyme complexation, it is necessary to establish both the loci of binding sites and the nature or types of interactions involved. In a recent paper Giacin and Gilbert (10) presented data which established that collagen-enzyme complexation involves regio-specific binding of the enzyme within the crystalline domain of the collagen microstructure. Further, the contribution of ionic interactions involving the E-amino group of lysyl side chains of collagen was determined by a chemical modification procedure (13). Such a procedure had been described by Grossberg and Pressman in studies on antigen-antibody binding (14).

In the latter studies, three reagents which modify lysyl E-amino residues in proteins were used to chemically modify collagen. Their effect on the binding of β-galactosidase (E. coli K_{12}) to membraneous collagen was evaluated by determination of the catalytic activity of the resultant collagen-enzyme complex. Modification of the lysyl residues was found to have a significant effect on enzyme binding, as the apparent specific activities of the resultant collagen-enzyme complexes decreased in proportion to the percent (mole percent) amino groups modified. These findings implied that,under the experimental conditions of pH etc. employed in these studies, lysyl E-amino groups function as active binding sites for enzyme protein-carrier protein binding.

In this paper, we present the effect of carbamylation on the binding of a purified source of β-galactosidase (E. coli K_{12}) to membraneous collagen. The effect of carbamylation on membraneous collagen has been analyzed in terms of the proportion of enzyme binding sites and the binding constants of the sites remaining. Binding constants were determined by the method of equilibrium sorption. Related studies with β-galactosidase isolated from A. niger are currently under investigation in our laboratory. Preliminary results of this study are also described.

Materials and Methods.

Enzymes. β-galactosidase (E. coli K₁₂) was obtained from Worthington Corporation, Freehold, New Jersey.

Collagen. Cattle-hide collagen was obtained from Devro, Inc. Somerville, New Jersey. The fibrous collagen was washed with 10% sodium chloride solution, followed by washing with distilled water. The collagen was freeze-dried and stored at 20°C.

Substrate. 0.15M Lactose in 0.02M sodium phosphate buffer at pH 7.0 was used as a substrate for β-galactosidase.

Purification of β-galactosidase. β-galactosidase (E. coli K₁₂) was purified by affinity chromatography. The inhibitor analog p-aminophenyl-β-D-galactopyranoside (obtained from Calbiochem. La Jolla, California 92037) was bound to agarose (obtained from Bio-Rad Laboratories, Richmond, California 94804) according to the method of Cuatrecasa (15). A modified stepwise elution procedure was employed in which the purified enzyme was eluted with 0.001 M Tris-buffer containing 0.01 M NaCl and 0.01 M MgCl₂, pH 7.5 (16). Disc electrophoresis analysis of the purified enzyme preparation showed one major band and a very faint zone of an impurity to be present. The purity of the enzyme was estimated to be approximately 90% by integration of the total absorption area of the electrophoresis gel scans. The gels were scanned at 280nm. Gels stained with Ceomasie blue were scanned at 550nm.

The specific activity of the enzyme was determined in 0.15M lactose solution (0.02M phosphate buffer, pH 7.0). The activity of the enzyme was expressed in terms of units of activity per mg of enzyme. A unit of activity was defined as a μ mole of glucose produced per minute. The soluble lactase activity following affinity chromatography purification was 37.1 units/mg. This represented a 4 fold increase in catalytic potency over the specific activity of the crude enzyme preparation (8.9 units/mg).

Preparation of Collagen Membranes. Collagen dispersion was prepared from hide collagen by dispersing freeze-dried collagen in aqueous lactic acid solution at pH 2.8. The collagen dispersion was cast to form a membrane (17). The thickness of the dried membrane was between 0.028 and 0.033 mm.

Preparation of Chemically Modified Collagen Membranes. Collagen membrane was prepared by the above procedure. The lysyl E-amino residues of collagen, in membrane form, were modified to varying degree by reaction with potassium cyanate. The collagen membrane was carbamylated by immersing preswollen films in 0.4M potassium cyanate solution, pH 8.5 for different periods of time. The carbamylation reaction was carried out at ambient temperature (approx. 20°C). The pH of the reaction mixture was monitored with a pH meter and the pH maintained at 8.5 by the addition of 0.05N

HCl solution. After treating the membrane for the specified time, the membrane was washed thoroughly with distilled water and the resulting membranes were swollen in 0.02M phosphate buffer, pH 7.0.

Preparation of Collagen-Enzyme Complexes. Collagen-enzyme complexes were prepared by the membrane impregnation method (9, 18). In this procedure, collagen membrane was formed and then impregnated with the purified enzyme. A solution containing 0.023μ mole/ml of enzyme in 0.02M phosphate buffer at pH 7.0 was employed as the impregnation bath. All enzyme impregnation reactions were carried out at 4°C for a twenty-four hour period. At the end of the impregnation period, the membrane was removed and dried at ambient temperature for eighteen hours. The resultant complex was washed with 0.02M phosphate buffer at pH 7.0. The activity of the collagen-enzyme complex was determined immediately after washing. The preparation of collagen-enzyme complexes from chemically modified collagen was carried out in a similar manner.

Equilibrium Sorption Procedure. The sorption (binding) of β-galactosidase by the various modified and unreacted collagen membranes was measured by the method of equilibrium sorption (18, 13). In this procedure, collagen membrane was impregnated with the purified enzyme (β-galactosidase, E. coli K_{12}) as a function of enzyme bath concentrations. Enzyme solutions containing 0.023, 0.012, 0.0074, 0.0047, 0.0031 and 0.0021 μmole/ml of enzyme in phosphate buffer (0.02M) at pH 7.0 were employed as the sorption baths. All sorption studies were carried out at 4°C for a twenty four hour period. Transient state sorption data established that the system had equilibrated within the twenty four hour period.

The initial and equilibrium concentration of soluble enzyme was determined spectrophotometrically by measurement of optical density at 280nm ($E^{1\%}_{1cm}$ = 20.9)(19). The concentration of sorbed enzyme was determined by difference. For the determination of soluble enzyme concentration, a 50 μl aliquot was withdrawn from the impregnation bath. The volume was adjusted to 5ml with 0.02M phosphate buffer, pH 7.0 and the optical density recorded at 280nm.

Appropriate control experiments were carried out, when both modified and unreacted collagen membrane was immersed in buffer solution (0.02M phosphate buffer, pH 7.0) and stored at 4°C for twenty four hours. Spectrophotometric analysis of the buffer media showed no absorbance from impurities and/or carrier protein at 280nm. Thus, the decrease in absorbance at 280 nm was a direct measure of the binding or sorption of enzyme to the collagen matrix.

Measurement of Lysyl Group Modification. The concentration of lysyl E-amino groups in collagen was determined by the method of Porter et al.(20). A fifteen mg sample of air dried collagen

film was preswollen in 10ml of 10% sodium bicarbonate solution.
Twenty ml of a 10% (wt/v) solution of dinitrofluorobenzene (DNFB)
in ethanol was added and the reaction mixture stirred for two
hours at ambient temperature. The membrane was removed from the
reaction mixture, washed with water, ethanol and diethylether
respectively and dried. The DNFB treated films were hydrolyzed
with 6N HCl for eighteen hours at 105°C in a capped test tube
flushed with nitrogen. After digestion, the hydrolyzate was extracted three(3) times with ether. The aqueous phase was diluted
to 100ml with 1N HCl and the absorbance at 363nm recorded. The
concentration of free lysyl residues in chemically modified collagen was carried out in an analogous manner.

Enzyme Assay Procedure. The catalytic potency of the immobilized β-galactosidase was determined in a plug flow reactor
(9). Glucose liberated by the catalytic activity of β-galactosidase on lactose was determined by the glucose oxidase-chromogen
method (21) with some modifications.
Chemical analysis for bound enzyme, following steady state
plug flow reactor operation, was based on the tryptophan content
of the complex. A modification of the method of Daiby et al.(22)
was employed.

Results and Discussion.

Enzyme Binding Capacity Determination. Two steps or stages
are involved in formation of a stable collagen/enzyme complex.
The initial step involves the sorption of enzyme from a free
enzyme solution by either a dispersion or a film of collagen (5,
9). The present study differs from previous investigations in
that in addition to determination of enzyme binding capacity of
collagen membrane by both chemical and apparent activity assays,
the initial binding relationship was also determined by classical sorption isotherm theory. A general agreement between the
two procedures was obtained. However, the specific relationships
differ in that the equilibrium sorption relationship is linear
with respect to lysyl group modification while the amount of
bound enzyme after the membrane has achieved steady state conditions in the reactor is non-linear.
Unless otherwise stated, the enzyme binding capacity of membraneous collagen was measured by the equilibrium sorption method.
In determining the contribution of ionic interactions, involving the E-amino group of lysyl residues of collagen, to the
binding of enzyme by collagen, a purified source of β-galactosidase was used. Since a spectrophotometric procedure was developed
to monitor the sorption of enzyme by membraneous collagen, it is
critical that the enzyme preparation be homogeneous and there be
no absorbance by impurities. Absorbance by impurities at 280nm
would interfere with quantitation of sorbed enzyme and lead to
erroneous results. Further, impurities present in the enzyme

preparation could compete with the enzyme for active binding sites on collagen, leading to inaccurate estimates for total binding site concentration (A_O) and binding constant (K_O) of the sites remaining.

Chemical Modification of Collagen- The Effect of Carbamylation on the Enzyme-Binding Capacity of Membraneous Collagen.

A. **Theoretical Considerations.** The mass action expression for the combination of enzyme protein with a homogeneous population of binding sites on the carrier protein, collagen, may be treated in terms of a Langmuirian isotherm, where the isotherm is expressed in the following form (18, 23):

$$b = K_O C (A_O - b) \qquad \text{or} \qquad (1)$$

$$\frac{1}{b} = \frac{1}{A_O} \left(\frac{1}{K_O C} \right) + \frac{1}{A_O} \qquad (2)$$

when A_O = total concentration of binding sites on the carrier protein; b = concentration of enzyme protein sorbed to carrier protein at equilibrium; C = concentration of soluble enzyme protein at equilibrium; and K_O = binding constant for the combining of enzyme to the binding sites on the carrier protein, collagen.

Some approximations can be made as to the concentration of active binding sites (A_O) on collagen and the binding constant (K_O) from the Langmuirian isotherm. Assuming that one enzyme molecule is bound per one binding site, Equation (2) predicts that when $\frac{1}{b}$ is plotted as a function of $1/C$, a straight line will be obtained and the intercept on the 1/b axis, that is at infinite free enzyme concentration, is equal to $1/A_O$. From A_O, the binding constant K_O can be obtained from the slope of the double reciprocal plot.

If the binding sites are heterogeneous, that is, more than one binding constant, the corresponding equation (3) is:

$$\frac{1}{b} = \frac{1}{A_O} \left(\frac{1}{K_O' C} \right)^a + \frac{1}{A_O} \qquad (3)$$

where (a) is the heterogeneity index and K_O' is the average binding constant. For a homogeneous group of sites, a=1 and $K_O' = K_\ominus$. Decreasing values of (a) correspond to an increasing heterogeneity of binding sites.

The free energy of binding (ΔG^O) of enzyme protein to carrier protein is also estimated from the relationship:

$$\Delta G^O = - RT \ln K_O \qquad (4)$$

B. **Effect of Carbamylation on Enzyme Binding.** Figure 1. shows the effectiveness of chemical modification (cambamylation) in reducing the enzyme binding activity of membraneous collagen, where the quantity of enzyme sorbed is plotted as a function of

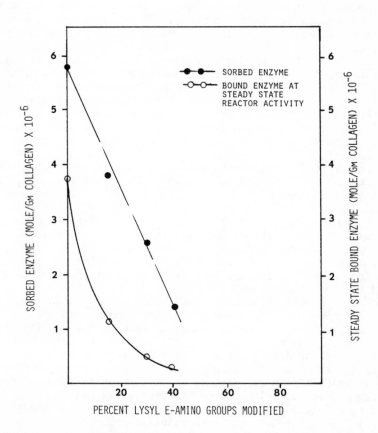

Figure 1.   The effect of chemical modification of lysyl E-amino groups on the sorption of β-galacatosidase (23 μmol/L) to collagen membrane.   Solution equilibrium and steady-state reactor conditions.

the percent (mole percent) E-amino groups modified. The equilibrium sorption studies described in Figure 1. were carried out at
a constant enzyme bath concentration (i.e. 0.023μ mole/ml). As
shown, chemical modification of the lysyl residues in collagen
had a marked effect on the binding or sorption of lactase (β-
galactosidase, E. coli K_{12}) to the modified membrane. A collagen
membrane in which 15% (mole%) of the available E-amino groups had
been converted to the carbamate derivative sorbed 35% less enzyme
than the unmodified control membrane. A 78% decrease in sorbed
enzyme was observed when 40% (mole %) of the lysyl E-amino groups
had been modified.

The decrease in enzyme sorption by the chemically modified
membrane implies that under these experimental conditions, the
lysyl E-amino groups function as principle receptor or binding
sites for enzyme protein (at least for E. coli β-galactosidase)
and that the complexation mechanism involves interaction of the
lysyl residues of collagen with enzyme amino acid chains as an
initial step in the formation of a stable network of physico-
chemical bonds.

The general relationship of enzyme binding to lysyl group
content is also found following steady state plug flow reactor
operation, as shown by the lower curve in Figure 1. The level of
bound enzyme reported in the lower curve (Figure 1.) is based on
the tryptophan content of the complex and represents the amount
of enzyme protein thermodynamically immobilized following steady
state plugflow reactor operation. In contrast to the linear relationship of lysine content to sorption equilibrium (Langmuirian
sorption isotherm) binding, the steady state on thermodynamically
bound species is nonlinearly related.

The relationship between lysyl group concentration and the
level of enzyme protein bound (following steady state plugflow
reactor operation) is further substantiated by the results shown
in Figure 2., where the catalytic activity of the collagen-enzyme
complex is plotted as a function of the percent E-amino groups
modified. As shown, a collagen-lactase complex prepared from a
film in which 15% (mole percent) of the available E-amino groups
had been converted to the carbamate derivative had an enzyme
activity 29% less than the complex prepared from the unmodified
control film. A 74% decrease in relative apparent activity was
observed for a complex prepared from a membrane in which 40 mole
percent of the lysyl groups had been modified.

As previously discussed, the decrease in enzyme binding capacity of the chemically modified membrane implies that the complexation mechanism involves ionic interactions of lysyl E-amino
groups of collagen with enzymic amino acid side chains as a principle step in the formation of a stable network of phyico-chemical
bonds. However, part of the effect of carbamylation could be due
to a reduction in the average binding constant (K_O) of all the
sites rather than the complete blocking of active sites.

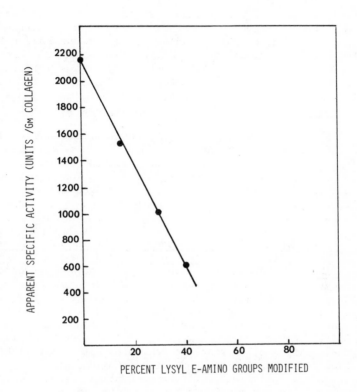

Figure 2. The effect of chemical modification of lysyl E-amino groups on the amount of bound enzyme at steady-state reactor conditions

Alternatively, the observed decrease in enzyme binding activity may be attributed to the destruction of a limited number of sites having a high binding constant. In order to ascertain how much loss of binding activity was due to reduction of K_O and how much was due to complete loss of active sites, sorption isotherms for β-galactosidase on chemically modified collagen membranes and for the corresponding unmodified control were determined at several enzyme concentrations. Collagen membranes in which 15, 30, and 40 mole percent of the lysyl E-amino groups had been blocked were used.

Sorption isotherms for β-galactosidase on the modified membranes and the corresponding unmodified control are presented in Figure 3. As shown, sorption followed a common Langmuirian-type isotherm, where the amount of β-galactosidase sorbed on the collagen membranes increased with increasing bath concentration, approaching a saturation value at higher enzyme concentrations. Similar results have been previously reported for lysozyme and lactase (24,25). In these earlier studies, the enzyme binding capacity of the membraneous collagen was evaluated by determination of the catalytic activity of the resultant collagen-enzyme complex. Our results from chemical modification of collagen and the affect on sorption or enzyme binding are consistent with the suggested mechanism of enzyme-collagen complexation involving a finite number of active binding sites on collagen and that the number of potential binding sites were decreased by chemical modification of the lysyl residues.

C. Thermodynamic Considerations: The Effect of Chemical Modification on Binding Constants and the Free Energy of Binding. In Figure 4., the sorption data are presented as the reciprocal of bound or sorbed enzymic protein (mole/gm collagen) versus the reciprocal of free or soluble sorbate concentration (mole/liter) under equilibrium conditions. The observed linear relationship obtained for the respective modified and unmodified membranes is suggestive of a heterogeneity index (a) of one. This implies that the combining sites on collagen are a homogeneous population, and the binding energies for the active sites are equivalent. If the assumption is made that one enzyme molecule is bound per one binding site, extrapolation of such curves to infinite free enzyme concentration, i.e. the ordinate intercept, provides a measure of the maximum level of enzyme binding and thus an estimated measure of the total concentration of binding sites (A_O).

In Table I, the resultant values for total binding site concentration (A_O), average binding constant (K_O) and the heterogeneity index (a) for the respective modified and unmodified membranes are presented. From Table I, it is seen that carbamylation of the lysyl residues of collagen results in the loss of enzyme binding sites (A_O) with little or no change in the average binding constant (K_O) for the remaining binding sites. This is

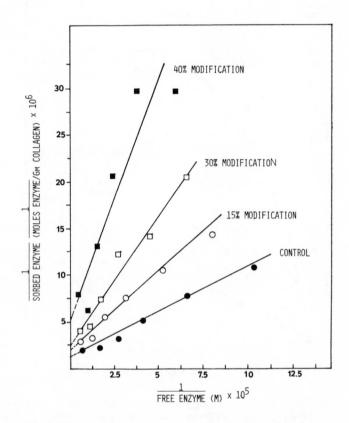

Figure 3. Sorption of β-galactosidase (E. coli K₁₂) by collagen preparations at different levels of lysyl group modification

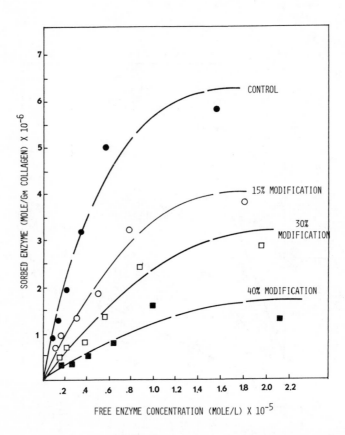

Figure 4. Sorption of β-galactosidase by collagen preparations (samples as in Figure 2) at different degrees of lysine content as a double reciprocal plot; for control, $A_o = 14 \times 10^{-6}$; for 15% modification, $A_o = 0.53 \times 10^{-6}$; for 30% modification, $A_o = 0.41 \times 10^{-6}$; for upper curve, $A_o = 0.18 \times 10^{-6}$ mol/g collagen, respectively

TABLE I

Total Available Binding Sites,

Binding Constants and Free Energy of Binding

Collagen Membrane (a) (Percent Modification)	Total Available Binding Sites (A_o) (Mole/Gm Collagen x 10^{-6})	Average Binding Constant (K_o) (ℓMole x 10^{-5})	Free Energy of Binding (ΔG^o) (Kcal/Mole)
Control	1.4	0.73	−6.2
15	0.53	1.7	−6.5
30	0.4	0.9	−6.3
40	0.2	1.2	−6.5

(a) Mole percent lysyl E-amino groups modified

consistent with the above discussion. Further, these binding
sites appear to be unaffected by distortion of the collagen
microstructure as a result of carbamylation of other loci within the tropocollagen molecule.

The estimated free energies of binding (ΔG) are also tabu-
lated in Table 1. As shown, the free energy of binding of enzyme
protein to the respective modified and unmodified collagen
membranes is equivalent. This is consistent with our previous
discussion on the heterogeneity index, which indicated that the
combining sites on collagen are a homogeneous population and the
binding energies for the sites are equivalent.

D. <u>Nature of Active Binding Sites of Collagen</u>. The negative values for the free energies of binding (ΔG) show that the
sorption process is energetically favorable and spontaneous.
Further, the estimated free energy value (-6.4 Kcal/mole) for
enzyme protein-carrier protein is similar to the free energy
values reported in the literature for protein-protein association. The value of -6.4 Kcal/mole is comparable to that for
association of the subunits of bovine liver glutamate dehydrogenase into active enzyme (26) and the binding of human hemoglobin α, β-dimer into the tetramer (27). Similar association
energies have been reported for the dimerization of α-chymo-
trypsin (28), insulin (29) and lysozyme (30). These findings are
summarized in Table II.

Based on x-ray crystallographic data, Fermi (31) concluded
that the binding of the α, β-dimer of human hemoglobin into the
tetramer involved 27 amino acid residues, including six (6)
hydrogen bonds, hydrophobic and Van der Waals interactions and
one possible salt bridge between the side chains of arginine and
glutamic acid. The crystalline structure of the insulin dimer
showed that 20 amino acid residues are involved in the dimerization (10 amino acid residues per monomer unit). These interactions included four (4) hydrogen bonds and Van der Waals forces
(32). Although Fermi (31) proposed the involvement of ionic
bonding in the formation of the hemoglobin tetramer, other
investigators (27, 33) have suggested that both the hemoglobin
tetramer and the insulin dimer are stabilized by hydrophobic
interactions. This assumption was based on the fact that these
moieties (i.e. hemoglobin tetramer and insulin dimer) were dissociated more readily by non-polar reagents.

By comparing the sorption data and the free energy values
for collagen-enzyme binding with the binding data for hemoglobin
and insulin, some approximations can be made as to the number
and the type of interactions involved in collagen-enzyme binding.
It is possible that the sorption or binding of β-galactosidase
to collagen may involve 20 to 30 amino acid residues. However,
our findings showed such an association to involve ionic interactions of lysyl E-amino groups of collagen with enzymic amino
acid side chains (i.e. carboxyl groups) as the primary step in
the binding of enzyme protein to the carrier protein,

TABLE II

Free Energy of Binding For
Various Protein-Protein Interactions

Proteins Involved	Nature of Binding	Free Energy (ΔG^O) Kcal/mole
Collagen-Lactase	Complexation	−6.4
Glutamate Dehydrogenase	Subunits $\overset{\rightarrow}{\leftarrow}$ active enzyme	−7.2
Human Hemoglobin	α,β-dimer $\overset{\rightarrow}{\leftarrow}$ tetramer	−7.1
α- Chymotrypsin	Monomer $\overset{\rightarrow}{\leftarrow}$ dimer	−5.7
Insulin	Monomer $\overset{\rightarrow}{\leftarrow}$ dimer	−5.3
Lysozyme	Monomer $\overset{\rightarrow}{\leftarrow}$ dimer	−3.9

collagen. The equivalency of the free energy for collagen-enzyme
sorption, irrespective of collagen treatment, further supports
the assumption that there are specific binding sites on the enzyme
molecule which form strong bonding interactions with lysyl residues of collagen.

As we have previously pointed out, a minimum of four such
ionic interactions per tropocollagen molecule would provide a
cumulative bond energy of approximately 100 Kcal/mole (10).

Enzyme Binding Mechanism.

Our findings on the effect of structural and chemical modification of collagen on enzyme binding permit certain general
conclusions regarding the mechanism of enzyme protein-carrier
protein binding.

1. The binding mechanism involves localization or regio-
specific binding of the enzyme within the crystalline region of
the collagen microstructure. The lysyl E-amino groups, localized
within the helical portion of the tropocollagen molecule, function
as principle receptor sites for enzyme binding. The complexation
mechanism involves ionic interactions of the lysyl E-amino group
with enzyme amino acid side chains (e.g. carboxyl groups) as a
principle step in the formation of a stable collagen-enzyme supra-
structure.

2. An essential step in the formation of a stable enzyme-
carrier protein complex involves drying of the impregnated membrane (17). This drying step is thought to reduce the intermolecular distance between enzyme amino acid side chains and
active binding sites on collagen, thereby facilitating the formation of a series of secondary physico-chemical bonds. It is this
cumulative binding which results in the formation of a stable
enzyme protein-carrier protein complex. The elimination of the
drying step results in excessive leaching of the enzyme from the
supporting matrix, following repeated contact with substrate
solution. The binding of enzymic protein to the carrier protein collagen therefore is assumed to involve two steps or stages.
The initial step involves interaction of enzymic amino acid side
chains with lysyl residues of collagen which affixes the enzyme
to the collagen matrix. This is followed by the formation of a
series of secondary non-covalent bonds during the drying step.

Conclusions.

The results of this study have delineated the mechanism of
enzyme protein-carrier protein binding in terms of the nature of
the functional group and/or groups within the helical region of
the collagen molecule involved in binding. Specifically the
role of lysyl E-amino groups of collagen has been established.
Further studies are in progress in our laboratory to determine
the equivalency of the binding mechanism involved in the immobilization of enzymic protein to collagen.

"Literature Cited".

1. Green, D. E., Murer, E., Hultin, H. O., Richardson, S. H., Salmon, B., Brierley, G. P. and Baum, H. Arch. Biochem. Biophys. (1965).112:635.
2. Arnold, H., and Pette, D. European J. Biochem., (1968).6:163.
3. Goldman, R., Goldstein, L., and Katchalski, E., Academic Press, (1971).
4. Goldman, R., Biochimie, (1973). 55:953.
5. Vieth, W. R., Gilbert, S. G., and Wang, S. S., Trans. N. Y. Acad. Sci., (1972). 34:454.
6. Wang, S. S., and Vieth, W. R. Biotechnol. Bioeng. (1973). 15:93.
7. Vieth, W. R., and Venkatasubramanian, K., Chemtech. Am.Chem. Soc., (1973).
8. Vieth, W. R., Amini, M. A., Constantinides, A., and Ludolph, R. A., I & EC Fundamentals., (1977). 16:82.
9. Lin, P. M., Giacin, J. R., Leeder, J. G., and Gilbert, S. G. J. Food Sci., (1976). 41:1056.
10. Giacin, J. R. and Gilbert, S. G. Protein Crosslinking, Part A, Plenum Publishing Co., (1977). P. 441.
11. Coulet, P. R., Julliard, J. H. and Gautheron, D. C.,Biotech and Bioengr. (1974). 16:1055.
12. Julliard, J. H., Godinot, C. and Gautheron, D. A., FEBS Letters, (1971). 14:185.
13. Luo, K. M., Giacin, J. R., and Gilbert, S. G. Presented 36th Annual IFT Meeting, Anaheim, CA, (1976).
14. Grossberg, A. L. and Pressman, D. Biochem. (1963). 2:90.
15. Cuatrecasa, P. J. Biol. Chem. (1970). 245:3059.
16. Luo, K. M. Ph.D. Thesis, Rutgers Univ. (1977).
17. Lieberman, E. R., Gilbert, S. G. and Shrinivasa, V., Trans. N. Y. Acad. Sci. (1972). 34:694.
18. Venkatasubramanian, K., Saini, R., and Vieth, W. R., J. Ferm. Technol. (1974). 52:268.
19. Nisonoff, A., and Pressman D.,J. Immunol. (1958). 80:417.
20. Porter, R. R., and Sanger, F. Biochem. J., (1948). 42:287.
21. Worthington Enzyme Manual, Worthington Biochem. Co., Freehold NJ., (1972).
22. Daiby, A. and Tsai, C. Anal. Biochem. (1975). 63:283.
23. Sips, R., J. Chem. Phys. (1949). 17:762.
24. Eskamani, A., Ph. D. Thesis, Rutgers Univ. (1972).
25. Venkatasubramanian, K., Vieth, W. R. and Wang, S. S., J. Ferm. Technol., (1972). 50:600.
26. Reisler, E. and Eisenberg, H. Biochem. (1971). 10:259
27. Herskovits, T. T. and Ibanez, V. S., Biochem. (1976). 15:5715.
28. Teller, D. C., Horbett, T. A., Richard, E. G. and Schachman, H. K., Ann. N. Y. Acad. Sci. (1969). 164:66.
29. Jeffery, P. D. and Coates, J. H., Biochem. (1966). 5:3820.
30. Banerjee, S. K., Pogolotti, A. Jr., and Rupley, J. A., J. Biol. Chem. (1975). 250:8260.

"Literature Cited" ctd.

31. Fermi, G.,J. Mol. Biol.,(1975). 97:237.
32. Blundell, T., Dodson, G., Hodgkin, D. and Mercola, D., Adv.
         Prot. Chem.,(1972). 26:279.
33. Chothia, C. and Janin, J. Nature. (1975). 256:705.

RECEIVED November 2, 1978.

Alkali-Induced Lysinoalanine Formation in Structurally Different Proteins

MENDEL FRIEDMAN

Western Regional Research Center, Science and Education Administration,
U.S. Department of Agriculture, Berkeley, CA 94710

Crosslinked amino acids have been identified in acid hydrolysates and enzyme digests of alkali-treated and heat-treated proteins. (Papers in references 1 and 2 cover this subject comprehen-
sively). One such crosslinking derivative, lysinoalanine, has
been found to cause histological (pathological) changes in rat
kidneys (3, 4). These observations cause concern about the nutritional quality and safety of alkali-treated foods. Chemical
changes that govern formation of unnatural amino acids during
alkali treatment of proteins need to be studied and explained,
and strategies to minimize or prevent these reactions need to be
developed.

In previous papers, we have (a) reviewed elimination reactions
of disulfide bonds in amino acids, peptides, and proteins under
the influence of alkali (5); (b) analyzed factors that may operate
during alkali-induced amino acid crosslinking and its prevention
(6); (c) demonstrated inhibitory effects of certain amino acids
and inorganic anions on lysinoalanine formation during alkali
treatment of casein, soy protein, wheat gluten, and wool and on
lanthionine formation in wool (7, 8, 9); (d) demonstrated that
protein acylation inhibits lysinoalanine formation in wheat gluten

0-8412-0478-0/79/47-092-225$05.00/0

and soy protein (8, 9); (e) studied the transformation of lysine
to lysinoalanine, and of cystine to lanthionine residues in
proteins and polyamino acids (10); and (f) examined effects of
lysine modification on chemical, nutritional, and functional
properties of proteins (11).

In this paper I report and discuss the susceptibilities of
alkali-labile amino acid residues in three proteins to degradation
as a function of pH.

EXPERIMENTAL

Commercial casein was obtained from International Casein
Corporation, San Francisco, California; commercial wheat gluten
and lactalbumin from United States Biochemical Corporation,
Cleveland, Ohio.

Alkali Treatments. The following procedure, illustrated with
casein, was also used with the other food proteins. A solution
or suspension of casein (usually 0.5 gram per 50 cc of solvent
or 1% w/v) in borate buffer of appropriate pH in a glass-stoppered
Erlenmeyer flask was placed in a 65°C water bath. After the
indicated time, the solution was dialyzed against 0.01N acetic
acid with frequent changes of water plus acetic acid for about
two days and then lyophilized.

Amino Acid Analyses. A weighed sample (about 5 mg) of protein
was hydrolyzed in 15 cc of 6N HCl in a commercial hydrolysis tube.
The tube was evacuated, placed in an acetone-dry ice bath, evacuated and refilled with nitrogen twice before being placed in an
oven at 110°C for 24 hours. The cooled hydrolysate was filtered
through a sintered disc funnel, evaporated to dryness at 40°C
with the aid of an aspirator, and the residue was twice suspended
in water and evaporated to dryness. Amino acid analysis of an
aliquot of the soluble hydrolysate was carried out on a Durrum
Amino Acid Analyzer, Model D-500 under the following conditions:
single column Moore-Stein ion-exchange chromatography method;
Resin, Durrum DC-4A; buffer pH, 3.25, 4.25, 7.90; photometer,
440 nm, 590 nm; column, 1.75 mm X 48 cm; analysis time, 105 min.
Norleucine was used as an internal standard.

In this system, lysinoalanine (LAL) is eluted just before
histidine. The color constant of LAL was determined with an
authentic sample purchased from Miles Laboratories.

Some typical results are shown in Figures 1-4.

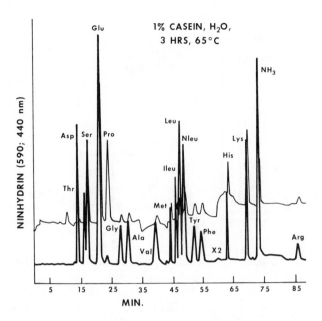

Figure 1.   *Amino acid analysis of a hydrolysate of casein heated in water*

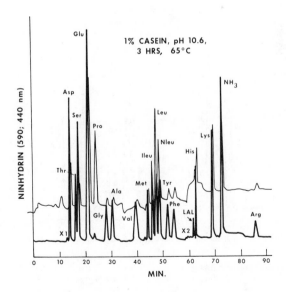

Figure 2.   *Amino acid analysis of a hydrolysate of casein heated in a pH 10.6 buffer. Note lysinoalanine peak.*

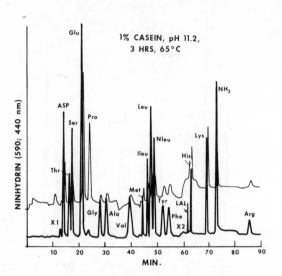

Figure 3.   Amino acid analysis of a hydrolysate of casein heated in a pH 11.2
buffer. Note lysinoalanine peak.

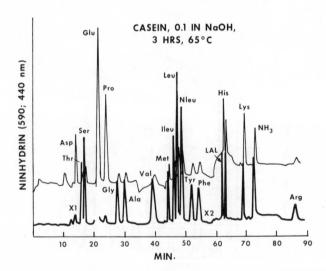

Figure 4.   Amino acid analysis of a hydrolysate of casein heated in a 0.1N NaOH
solution. Note lysinoalanine peak.

RESULTS AND DISCUSSION

The amino acid composition of alkali-treated casein, lactalbumin, and wheat gluten are given in Tables I–III. The results
show that the following amino acids are destroyed to various
extents under basic conditions: threonine, serine, cystine, lysine,
and arginine, and possibly also tyrosine and histidine. The losses
of these amino acids is accompanied by the appearance of lysinoalanine and other ninhydrin-positive compounds.

Inspection of the Tables reveals several interesting points.
First, loss of lysine appears to level off or go through a minimum with increasing pH. A possible explanation is that lysinoalanine is destroyed (besides being formed) during alkali treatment,
regenerating lysine. An analogous regeneration of lysine has
been shown to occur when ε-cyanoethyl derivatives of lysine are
subjected to alkaline conditions (12). This possibility is
supported by the following observations. (a) Time studies show
that lysinoalanine formation appears to level off after about
one hour when lactalbumin or soy protein is exposed to 1N NaOH
at 65°C up to 8 hours (9). (b) Exposure of free- and protein-
bound lysinoalanine to alkaline conditions appears not to always
give quantitative recovery of lysinoalanine (13).

Comparison of lysinoalanine values for wheat gluten, casein
and lactalbumin   treated at various pH's shows large differences
in the amounts of lysinoalanine formed in the three proteins.
For example, the respective values at pH 10.6 are 0.262, 0.494,
and 1.04 mole per cent (ratio of about 1:2:4); at pH 11.2 the
values are 0.420, 0.780, and 1.52 mole per cent; and at pH 12.5
(pH of 1% protein solution in 0.1N NaOH), the respective values
are 0.762, 0.780, and 2.62 mole per cent. (Note that the value of
casein approaches that of gluten at this pH). The observed
differences in lysinoalanine content of the three proteins at
different pH values are not surprising since the amino acid composition, sequence, protein conformation, molecular weights of
protein chains, initial formation of intra- versus intermolecular
crosslinks may all influence the chemical reactivity of
a particular protein with alkali. Therefore, it is not surprising
to find differences in lysinoalanine content in different proteins
treated under similar conditions. These observations could have
practical benefits since, for example, the lower lysinoalanine
content of casein compared to lactalbumin treated under the same
conditions suggests that casein is preferable to lactalbumin
in foods requiring alkali-treatment.

The postulated mechanism of lysinoalanine formation (Figure 5)
is at least a two-step process. First, hydroxide ion-catalyzed
elimination reactions of serine, threonine, and cystine (and to

TABLE I

Effect of pH on amino acid composition of wheat gluten.
Conditions: 1% wheat gluten;   65°C;   3 hours.
Numbers are mole (residue) per cent for each amino acid.

Amino Acid	Control	pH 9.6	10.6	11.2	12.5	13.9
ASP	3.20	3.26	3.26	3.15	3.57	2.96
THR	3.19	3.10	3.05	3.01	2.67	1.20
SER	6.81	6.75	6.64	6.55	5.30	2.24
ALA	3.91	3.95	4.07	3.86	4.40	3.78
CYS	0.976	0.691	0.00	0.00	0.00	0.00
MET	1.35	1.25	1.18	1.33	1.68	1.17
TYR	2.49	2.55	2.43	2.49	2.43	1.85
PHE	4.35	4.29	4.52	4.35	4.29	4.93
LAL	0.00	0.00	0.262	0.420	0.762	0.884
HIS	1.87	1.80	1.83	1.78	1.74	1.71
LYS	1.33	1.40	1.16	0.963	0.945	0.948
ARG	2.75	2.70	2.66	2.68	2.61	1.79

TABLE II

Effect of pH on amino acid composition of <u>casein</u>. Conditions:
1% commercial casein; 65°C; 3 hours. Numbers are mole per cent
for each amino acid.

Amino Acid	Controls (in triplicate)			pH		
	No 1	No 2	No 3	10.6	11.2	12.5
ASP	6.85	6.81	6.85	6.93	6.96	7.04
THR	4.61	4.58	4.44	4.40	4.44	3.90
SER	7.12	7.07	7.27	7.09	6.94	4.60
GLU	18.84	18.76	19.18	19.14	19.46	20.83
PRO	11.97	12.08	12.28	12.20	12.30	12.56
GLY	3.18	3.17	3.05	3.05	3.08	3.44
ALA	4.34	4.36	4.25	4.28	4.20	4.49
VAL	6.81	6.89	6.71	6.66	6.60	6.88
MET	2.43	2.46	2.43	2.36	2.00	2.74
ILEU	4.73	4.80	4.72	4.71	4.78	4.70
LEU	9.37	9.42	9.21	9.28	9.48	9.11
TYR	3.87	3.87	3.86	3.89	3.82	3.88
PHE	4.02	3.96	3.95	4.04	4.09	4.21
LAL	0.00	0.00	0.00	0.494	0.780	2.43
HIS	2.40	2.39	2.42	2.48	2.43	2.39
LYS	6.86	6.76	6.89	6.46	6.03	4.48
ARG	2.59	2.60	2.49	2.53	2.50	2.21

TABLE III

Effect of pH on amino acid composition of lactalbumin. Conditions:
1% lactalbumin: 65°C, 3 hours. Numbers are mole (residue) percent
values of the total accounted for.

| Amino Acid | Control | pH | | | | |
		9.60	10.60	11.20	12.50	13.90
ASP	12.02	11.68	11.34	12.31	12.14	15.55
THR	6.01	6.04	6.00	5.95	5.33	2.75
SER	6.18	6.11	6.17	6.08	5.44	2.26
ALA	7.40	7.78	7.74	7.50	7.97	8.22
CYS	0.952	0.558	0.190	0.00	0.00	0.00
MET	1.84	1.96	2.12	2.06	1.92	2.14
TYR	2.59	2.56	2.75	2.70	2.90	2.42
PHE	2.93	2.91	3.04	3.00	3.21	2.72
LAL	0.00	0.255	1.04	1.52	2.62	3.87
HIS	1.66	1.60	1.67	1.64	1.51	1.20
LYS	8.94	8.57	7.47	7.45	6.48	7.19
ARG	2.14	2.22	2.15	2.12	1.61	1.32

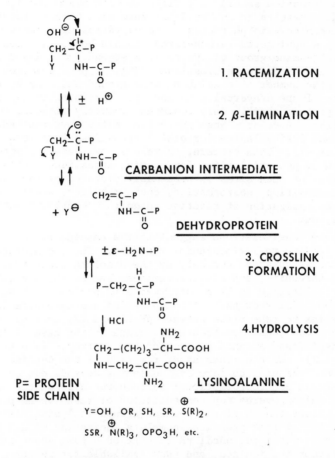

1. RACEMIZATION

2. β-ELIMINATION

CARBANION INTERMEDIATE

DEHYDROPROTEIN

3. CROSSLINK
FORMATION

4.HYDROLYSIS

LYSINOALANINE

P= PROTEIN
SIDE CHAIN

Figure 5.   Transformation of reactive protein side chains to lysinoalanine side chains via elimination and crosslinking formation.

Hydroxide ion abstracts an acidic hydrogen atom (proton) from an α-carbon atom of an amino acid residue to form an intermediate carbanion. The carbanion, which has lost the original asymmetry of the amino acid residue, can either recombine with a proton to reform a racemized residue in the original amino acid side chain or undergo the indicated elimination to form a dehydroalanine side chain. The dehydroalanine then combines with an ε-amino group of a lysine side chain to form a crosslinked protein which on hydrolysis yields free lysinoalanine (6, 9).

a lesser extent probably also cysteine) give rise to a dehydroalanine intermediate. Since such elimination reactions are second-
order reactions that depend directly on the concentration of both
hydroxide ion and susceptible amino acid, the extent of lysinoalanine formation should vary directly with hydroxide ion concentration. Results in Tables I-III show that this is indeed the
case within certain pH ranges. The dehydroalanine residue, which
contains a conjugated carbon-carbon double bond, then reacts
with the ε-amino group of lysine in a second, second-order step
to form a lysinoalanine crosslink. This step is governed not
only by the number of available amino groups but also by the
location of the dehydroalanine and amino group potential partners
in the protein chain. Only residues favorably situated to form
crosslinks can do so. When convenient sites have reacted,
additional lysinoalanine (or other) crosslinks form less readily
or not at all. Each protein, therefore, may have a limited
fraction of potential sites for forming crosslinked residues.
The number of such sites is presumably dictated by the protein's
size, composition, conformation, chain mobility, steric factors,
extent of ionization of reactive amino (or other) nucleophilic
centers, etc.

These considerations suggest that a cascade of reactions
occurs leading to lysinoalanine residues. Thus, dehydroalanine
formation is governed not only by the absolute concentration of
serine, threonine, and cystine residues but by their relative
susceptibilities to base-catalyzed eliminations. Thus, results
in Tables I-III show that although serine and threonine destruction begins to take place between pH 11 and 12, cystine residues
are much more sensitive to alkali, since in the case of lactalbumin, significant amounts of cystine are destroyed even at pH 9.6
(Table III). On the other hand, reaction of the ε-amino groups
with dehydroalanine to form lysinoalanine depends not only on the
cited steric and conformational factors but also on the pH of the
medium, which governs the concentration of reactive nonprotonated
amine. Since the pK of the ε-amino groups of lysine residues
is near 10 for most proteins, complete ionization of all amino
groups does not occur until pH 12. At pH 9 only about 10% of the
amino groups are ionized, and thus available for reaction (Cf. 14)
(All of the amino groups can eventually react, however, since
additional amino groups are formed by dissociation of the proto-
nated ammonium ions as the nonprotonated amino groups are used up)

These results, therefore, imply that the extent of lysinoalanine formation may vary from protein to protein. Factors that
favor or minimize these reactions need to be studied seprately
with each proteins.

ABSTRACT

Lysinoalanine formation in casein, lactalbumin, and wheat
gluten was measured at 65°C at various pH's for 3 hours. Factors
that control the extent of formation of the unnatural amino acid
lysinoalanine during food processing and thus the degree of
crosslinking in structurally different proteins are discussed.

LITERATURE CITED

1. Friedman, M., Ed., (1977). "Protein Crosslinking: Nutritional and Medical Consequences", Plenum Press, New York, 740 pages.

2. Friedman, M., Ed., (1977). "Protein Crosslinking: Biochemical and Molecular Aspects", Plenum Press, New York, 760 pages.

3. Woodard, C.J., Short, D.D., Alvarez, M.R. and Reyniers, J. (1975). Biological effects of N-ε-(DL-2-amino-2-carboxyethyl)-L-lysine, lysinoalanine. In "Protein Nutritional Quality of Foods and Feeds", M.Friedman, Ed., Marcel Dekker, New York, Part 1, pp. 595-616.

4. Gould, D. H. and MacGregor, J. T. (1977). Biological effects of alkali-treated protein and lysinoalanine: an overview. Reference 1, pp. 29-48.

5. Friedman, M. (1973). "Chemistry and Biochemistry of the Sulfhydryl Group in Amino Acids, Peptides, and Proteins", Pergamon Press, Oxford, England and Elmsford, New York, Chapter 5.

6. Friedman, M. (1977). Crosslinking amino acids--stereochemistry and nomenclature. Reference 1, pp. 1-27.

7. Finley, J. W.,Snow, J. T., Johnston, P. H. and Friedman, M. (1977). Reference 1, pp. 85-92.

8. Friedman, M. (1978). Wheat gluten-alkali reactions. In "Proceedings of the 10th National Conference on Wheat Utilization Research", U. S. Department of Agriculture, Science and Education Administration, Western Regional Research Center, Berkeley, California 94710, ARM-W-4, pp. 81-100.

9. Friedman, M. (1978). Inhibition of lysinoalanine synthesis by protein acylation. In "Nutritional Improvement of Food and Feed Proteins", M. Friedman, Ed., Plenum Press, New York, pp. 613-648.

10. Friedman, M., Finley, J. W. and Yeh, Lai-Sue (1977). Reactions of proteins with dehydroalanine. Reference 1, pp. 213-224.

11. Friedman, M. (1977). Effects of lysine modification on chemical, physical, nutritive, and functional properties of proteins. In "Food Proteins", J. R. Whitaker and S. R. Tannenbaum, Eds., Avi, Westport, Connecticut, pp. 446-483.

12. Cavins, J. F. and Friedman, M. (1967). New amino acids derived from reactions of ε-amino groups with α,β-unsaturated compounds. Biochemistry, 6, 3766-3770.

13. Friedman, M. and Noma, A. T., manuscript in preparation.

14. Friedman, M. and Williams, L. D. (1977). A mathematical analysis of consecutive, competitive reactions of protein amino groups. Reference 1, pp. 299-319.

RECEIVED November 3, 1978.

INDEX

INDEX